Linear Dynamical Systems

Mircea D. Grigoriu

Linear Dynamical Systems

Springer

Mircea D. Grigoriu
Department of Civil & Environmental
Engineering
Cornell University
Ithaca, NY, USA

ISBN 978-3-030-64554-0 ISBN 978-3-030-64552-6 (eBook)
https://doi.org/10.1007/978-3-030-64552-6

This Springer imprint is published by the registered company Springer Nature Switzerland AG
The registered company address is: Gewerbestrasse 11, 6330 Cham, Switzerland

To the class of 2020

Contents

Chapter 1
Introduction

This book is designed for seniors, first year graduate students and engineers working on the analysis and design of mechanical/structural/aerospace systems subjected to dynamic actions, e.g., wind, earthquakes, aerodynamic forces, road roughness and other inputs. It provides a comprehensive rigorous discussion on the dynamics of linear systems in clear, concise terms.

The main reason for a new book on linear dynamical systems is that current popular books on the subject devote numerous pages to simple dynamic systems of limited practical interest, e.g., cars without shock absorbers, before considering realistic systems, e.g., functional cars that have shock absorbers. Moreover, this bottom-up approach results in books that are overwhelming due to shear size and has been confusing to many Cornell students since they view these simplified dynamical systems as distinct problems which require different tools for solution. In contrast, this book deals, first, with realistic systems. Simple dynamical systems become special cases of the general formulation.

The inspiration for this book came from teaching a dynamics course online to Cornell University students during the 2020 Covid-19 pandemic. Out of necessity, I organized my lectures over the years into this self-contained, comprehensive work intended to be accessible to university students and practitioners. The impact of the pandemic made it apparent that online teaching would become more and more the norm for delivering lectures to college students and, it will give access to engineers in the field that was not available to them before in traditional, brick-n-mortar educational settings.

Notable attributes of this book are: it covers all relevant topics on the dynamics of linear systems in a succinct and rigorous manner; develops general methods for analyzing dynamical systems and views solutions of simplified dynamical systems as special cases of general formulations so that they cannot be viewed as distinct problems; discusses less familiar mathematical tools used in some derivations in appendices; uses numerical examples to further clarify theoretical considerations; and, lastly, illustrates practical implementation by MATLAB functions.

M. D. Grigoriu, *Linear Dynamical Systems*,
https://doi.org/10.1007/978-3-030-64552-6_1

The book has three parts. The first part (Chap. 2) is on single degree of freedom (SDOF) systems. The third part (Chaps. 4 and 5) is on the multi-degree of freedom (MDOF) and continuous systems. The insertion of Chap. 3 on the eigenvalue problem between developments on SDOF systems of Chap. 2 and MDOF and continuous systems of Chaps. 4 and 5 is intentional for the following reason. This nonstandard approach has been successful with the Cornell students, since, if one masters the theory of SDOF systems and of the eigenvalue problem, he/she will be capable to reconstruct developments in the third part of the book (Chaps. 4 and 5) by viewing the displacement functions of MDOF/continuous systems as elements of the vector spaces spanned by the eigenvectors/eigenfunctions of these systems whose components are defined by equations of the types satisfied by SDOF systems.

The number of degrees of freedom is equal to the number of parameters (functions of time) needed to specify the position of the masses of a system at all times. Consider a massless blade fixed at one end and free at the other end with a concentrated point mass, which vibrates in the plane of the paper. If the blade deformation is small, the position of its concentrated mass (the only mass in the system) can be described with good approximation by a single function of time, e.g., the distance of the concentrated mass to the undeformed blade, see Fig. 1.1a. The system has a single degree of freedom. If the blade deformation is large, two functions are required to describe the position of the concentrated mass, e.g., the distance of the concentrated mass to the undeformed blade and its distance to the fixed end of the blade, see Fig. 1.1b. The system has two degrees of freedom. We refer to systems that require $n \geq 2$ functions of time to describe the position of their masses at all times as multi-degree of freedom systems with n degrees of freedom. A blade with $n > 1$ point masses undergoing small deformation is another example of MDOF system with n degrees of freedom, see the system in Fig. 1.1c which has 3 masses so that $n = 3$. If the assumption that the blade is massless is removed, the blade has masses at all spatial locations so that an infinite number of functions of time is needed to describe the position of the system masses at all times, see

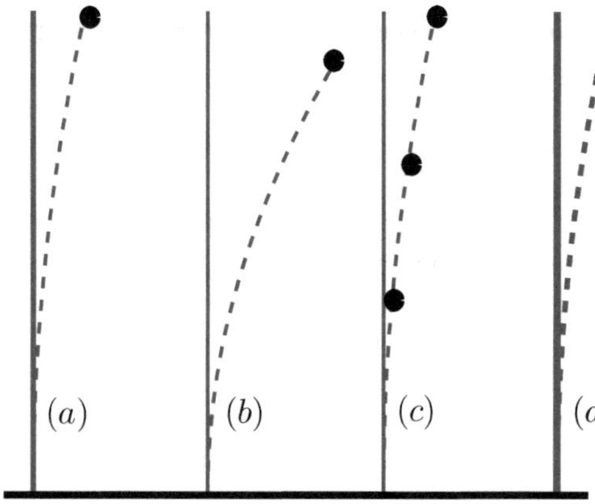

Fig. 1.1 Number n of degrees of freedom: (**a**) $n = 1$, (**b**) $n = 2$, (**c**) $n = 3$, and (**d**) $n = \infty$

Fig. 1.1d. The system has an infinite number of degrees of freedom and is referred to as continuous system or system with distributed mass.

As stated, the developments on MDOF/continuous systems (Chaps. 4 and 5) are based on methods for solving SDOF systems (Chap. 2) and representations of the deformations of MDOF/continuous systems in eigenvector/eigenfunction coordinates developed in Chap. 3 and involve the following three steps. First, the displacement vectors/functions of MDOF/continuous systems are viewed as elements of the linear spaces spanned by the eigenvectors/eigenfunctions of these systems. The methods of Chap. 3 are employed to construct eigenvector/eigenfunction coordinates. The displacement functions are completely defined by their projections on these coordinates, which are finite for MDOF systems and (countable) infinite for continuous systems. Second, differential equations are developed for the projections of the displacement vectors/functions of MDOF/continuous systems on their eigenvectors/eigenfunctions. These equations are uncoupled and have the structure of the equations of motion for SDOF systems. They can be solved one-by-one by using the methods developed in Chap. 2. Third, the displacement vectors/functions of MDOF/continuous systems are assembled from their representations in the eigenvector/eigenfunction coordinates and their projections of these coordinates.

The problems at the ends of the chapters are intended to facilitate the understanding of basic concepts and tools for solutions. The instructor is encouraged to develop and discuss, in addition to these problems, field-specific problems.

Finally, I would like to acknowledge the significant contributions of the many students enrolled in my class on dynamics and the teaching assistants participating in this class for their questions and comments. Their input was essential to the completion of this work and is highly appreciated.

Chapter 2
Single Degree of Freedom (SDOF) Systems

We review briefly the second law of Newton and its impulse-momentum and work-energy versions and use them to write equations of motion for single degree of freedom systems. These equations are solved by analysis in the time and frequency domains. Analytical and numerical methods are used for solution. Multi-degree of freedom and continuous systems are considered in Chaps. 4 and 5.

2.1 Newton's Second Law

Consider a material point with mass m which is subjected to a, generally, time-dependent force f. The point moves with velocity v along the real line according to the ordinary differential equation

$$\frac{d}{dt}\big(m\,v\big) = f \quad \text{or, equivalently, } d\,\big(m\,v\big) = f\,dt, \quad \text{(Newton's law (NL)).} \tag{2.1}$$

It is common to denote the point position, velocity, and acceleration by x or $x(t)$, $\dot{x}$, $\dot{x}(t)$, v, or $v(t)$, and $\ddot{x}$, $\ddot{x}(t)$, a, or $a(t)$. Similarly, the time argument of the mass and force may or may not be written so that they are denoted by m or $m(t)$ and f or $f(t)$. If the mass is time-invariant, the Newton law becomes

$$m\,a(t) = f(t) \quad \text{or, with the above convention, } m\,a = f. \tag{2.2}$$

The ***impulse-momentum*** (IM) and the ***work-energy*** (WE) relationships result from Newton's law (NL) by elementary manipulations. They are useful in applications since the solutions of some problem by these relationships are more efficient than that by Newton's law.

M. D. Grigoriu, *Linear Dynamical Systems*,
https://doi.org/10.1007/978-3-030-64552-6_2

Impulse-Momentum: The integral $\int_{t_1}^{t_2} d\,(mv) = \int_{t_1}^{t_2} f\,dt$ of Eq. 2.1 over a time interval (t_1, t_2) gives

$$m(t_2)\,v(t_2) - m(t_1)\,v(t_1) = \int_{t_1}^{t_2} f(t)\,dt. \tag{2.3}$$

The left and right sides of this equation are the change of momentum during the time interval (t_1, t_2) and the area under the force graph in (t_1, t_2). If the mass is time invariant, we have $m\,v(t_2) - m\,v(t_1) = \int_{t_1}^{t_2} f(t)\,dt$.

Work-Energy: Suppose that the mass is time invariant so that Newton's law has the form in Eq. 2.2. This equation becomes $m\,\ddot{x}(t)\,\dot{x}(t)\,dt = f(t)\,\dot{x}(t)\,dt$ by multiplication to $dx(t) = \dot{x}(t)\,dt$ or, equivalently,

$$d\left(\frac{1}{2}\,m\,\dot{x}(t)^2\right) = f(t)\,dx(t). \tag{2.4}$$

The left and right sides of this equations are the variation of the kinetic energy during the time dt and the work of the force over the space increment $dx(t)$. The left and right sides of the integral

$$\frac{1}{2}\,m\,v(t_2)^2 - \frac{1}{2}\,m\,v(t_1)^2 = \int_{t_1}^{t_2} f(t)\,dx(t) \tag{2.5}$$

of Eq. 2.4 over the time interval (t_1, t_2) represent the change in the kinetic energy and the work of the applied force during (t_1, t_2).

The following example illustrates the applications of the above formulations and their efficiency depending on the problem under consideration. We also note that equations of motion are meaningful if and only if a system of coordinates is specified.

Example 2.1 An object with mass m is dropped from elevation h above a linear spring with stiffness k, see Fig. 2.1. It is assumed that the spring has no mass and deforms along a straight vertical line and that the mass is attached to the spring following initial contact. Our objective is to find the maximum displacement δ of the spring caused by the falling mass. We have to deal with two distinct systems, a free mass from 1 to 2 and a mass supported by a spring from 2 to 3.

System 1–2 Consider the system from 1 to 2 with the system of coordinates in the figure. The applied force $f = m\,g$ is the weight of the mass, where g denotes the constant of gravity. NL gives $m\,\ddot{x} = m\,g$ or $\ddot{x} = g$ so that $\dot{x}(t) = g\,t + c_1$ and $x(t) = g\,t^2/2 + c_1\,t + c_2$. The initial conditions are $x(0) = 0$ and $\dot{x}(0) = 0$ in our system of coordinates so that $c_1 = c_2 = 0$ and $x(t) = g\,t^2/2$. The time $\bar{t}$ at which the mass reaches the top of the spring results from $h = g\,\bar{t}^2/2$ and is $\bar{t} = \sqrt{2\,h/g}$. The impact velocity is $\dot{x}(\bar{t}) = g\,\bar{t} = \sqrt{2\,g\,h}$.

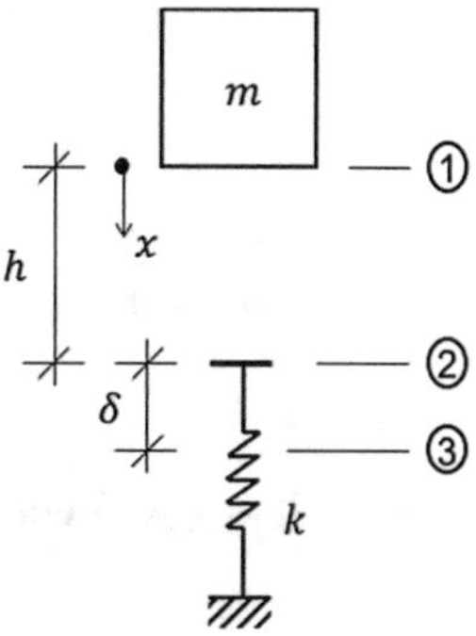

Fig. 2.1 Mass m dropped from elevation h on a spring with stiffness k

The solutions by the IM relationship deliver the impact velocity with similar calculations. We have $m\,v(t) - m\,v(0) = \int_0^t m\,g\,ds = m\,g\,t$ by IM so that $v(t) = g\,t$ since $v(0) = 0$. This gives $x(t) = g\,t^2/2 + c$ which becomes $x(t) = g\,t^2/2$ since $x(0) = 0$. The rest of calculations are as in the solution by the NL.

The solution by the WE relationship is slightly more efficient. We have $(1/2)\,m\,v(\bar{t})^2 - (1/2)\,m\,v(0)^2 = \big(m\,g\big)\,h$ so that the impact velocity is $v(\bar{t}) = \sqrt{2\,g\,h}$ since $v(0) = 0$.

System 2–3 Consider the mass-spring system and a different system of coordinates, the coordinate x' which measures the deformation of the spring relative to its undeformed position so that $x = x' + h$. This new coordinate is used for convenience. Any other coordinate can be used. The solutions by the NL and IM momentum relationship involve lengthy calculations using tools which will be discussed later in this chapter. On the other hand the solution by the WE relationship is much simpler. It is based on the observations that (1) the mass velocity changes sign at the time the spring reaches its maximum deformation δ and (2) the force acting on the mass at an arbitrary deformation x' is $m\,g - k\,x'$, where $m\,g$ is the weight of the mass and $-k\,x'$ denotes the response of the spring which opposes deformation. Accordingly, the WE relationship has the form

$$(1/2)\,m\,v_3^2 - (1/2)\,m\,v_2^2 = \int_0^{\delta} \big(m\,g - k\,x'\big)\,dx',$$

where $v_2 = \sqrt{2\,g\,h}$ is the impact velocity and $v_3 = 0$ is the velocity at the maximum deformation. This gives the algebraic equation $-(1/2)\,\big(2\,g\,h\big) = m\,g\,\delta - k\,\delta^2/2$ for δ which has the solutions $\delta_{1,2} = \delta_{\,\mathrm{st}} \pm \sqrt{\delta_{\mathrm{st}}^2 + 2\,h\,\delta_{\mathrm{st}}}$, where $\delta_{\mathrm{st}} = m\,g/k$ denotes the static deformation under the weight of the mass. Since the solution with minus is physically inadmissible (it suggests that the spring stretches under compression load), we conclude that the maximum deformation of the spring is

$$\delta = \delta_{\mathrm{st}} \left(1 + \sqrt{1 + \frac{2h}{\delta_{\mathrm{st}}}} \right) > \delta_{\mathrm{st}}. \tag{2.6}$$

Note that δ is twice as large as the statistical deformation δ_{st} if the spring is loaded suddenly ($h = 0$) rather than statically.

2.2 Physical System

Consider the physical system in Fig. 2.2 consisting of a mass m which can slide on a frictionless surface and is connected to a linear elastic spring with stiffness $k > 0$ and a linear viscous damper with parameter $c > 0$. The system is subjected to a forcing function $f(t)$ applied to the mass (Not shown in the figure!). Denote by $x(t)$ the position of the mass m at time $t \geq 0$ with respect to the x-coordinate in the figure which measures the position of the mass with respect to its location corresponding to the undeformed spring.

At any time $t \geq 0$, the mass is subjected to three forces: the applied force, the spring restoring force, which is proportional to the deformation $x(t)$, and the damper restoring force, which is proportional to the velocity $\dot{x}(t)$, for linear springs and viscous dampers as considered in our discussion. The signs of these forces are determined by the system of coordinates. For the x-coordinate in Fig. 2.2, the applied force $f(t)$ is positive if it acts in the positive direction of the x-coordinate. The spring and damper restoring forces, $-k\,x(t)$ and $-c\,\dot{x}(t)$, are negative if $x(t) > 0$ and $\dot{x}(t) > 0$, see illustration in the figure. Note that the signs of the restoring forces are determined by those of the oscillator displacement and velocity. For example, the spring and damper restoring forces are negative and positive if $x(t) > 0$ and $\dot{x}(t) < 0$.

Any other system of coordinates can be used. For example, suppose that the mass position is measured from a point left to the current origin at distance $a > 0$. Denote by $y(t)$ the position of the mass relative to this origin. Since $y(t) = a + x(t)$, the elastic and damping forces in this system of coordinates are $-k\,\big(y(t) - a\big)$ and $-c\,\dot{y}(t)$. As expected, the solutions in the two system of coordinates coincide. We will revisit this statement shortly.

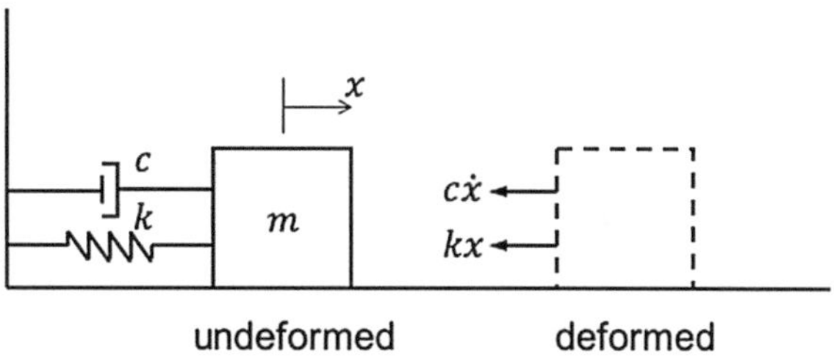

Fig. 2.2 Physical model of SDOF systems

The system in Fig. 2.2 has a single degree of freedom since the position of its mass is completely defined at all times by, e.g., the function $x(t)$ in the x-coordinate or the function $y(t)$ in the y-coordinate.

2.3 Equation of Motion

Newton's law (mass times acceleration equals the applied actions) for the mass in Fig. 2.2 in the x-coordinate gives the following differential equation:

$$m\,\ddot{x}(t) = f(t) - k\,x(t) - c\,\dot{x}(t) \tag{2.7}$$

for the displacement function $x(t)$. It is assumed that the forcing function $f(t)$ is positive, i.e., it acts from left to right in the positive direction of the x-coordinate. The displacement function $x(t)$ satisfies an ordinary differential equation of order 2 with constant coefficients since k and c are assumed to be constant. The single and double dots denote first- and second-order time derivatives. With the notations $\omega^2 = k/m$ and $2\,\zeta\,\omega = c/m$, the above equation takes the form

$$\ddot{x}(t) + 2\,\zeta\,\omega\,\dot{x}(t) + \omega^2\,x(t) = f(t)/m. \tag{2.8}$$

The notation $\omega^2 = k/m$ makes sense since k and c are positive. We will see that the parameters $\omega > 0$ and $\zeta \geq 0$ have precise physical meaning. The solution of Eqs. 2.7 and 2.8 requires initial conditions (ICs) which consists of the displacement $x(0) = x_0$ and the velocity $\dot{x}(0) = \dot{x}_0$ at the initial time, where x_0 and $\dot{x}_0$ are specified numbers.

The differential equations 2.7 and 2.8 describe the motion of most general SDOF systems, i.e., the motion of ***damped oscillators*** in ***forced vibration*** if $f(t) \neq 0$. Special cases of these equations of interest in applications are considered extensively in this chapter. For example, the SDOF system is said to be in ***free vibration*** if there is no applied force, i.e., $f(t) = 0$, and its vibrations are ***undamped*** if $c = 0$ or, equivalently, $\zeta = 0$.

The solution of Eqs. 2.7 and 2.8 can be obtained by analysis in the ***time domain*** and the ***frequency domain***. We begin with the time domain analysis and present several approaches for calculating the displacement function $x(t)$. The latter part of this chapter deals with the frequency domain analysis. This approach is particularly useful for SDOF systems which exhibit steady-state motion, i.e., the motions observed in some systems for large times. A formal definition of the steady-state response is given shortly.

2.4 Time Domain Analysis

Consider the equation of motion in the x-coordinate, i.e., Eq. 2.8. The ***general solution*** $x(t)$ of this equation is the sum

$$x(t) = x_h(t) + x_p(t) \tag{2.9}$$

of the general solution of the ***homogeneous equation*** $x_h(t)$, i.e., Eq. 2.8 with $f(t) = 0$, and a particular solution $x_p(t)$ of the ***inhomogeneous equation***, i.e., Eq. 2.8 with the actual input $f(t)$. The functions $x_h(t)$ and $x_p(t)$ are referred to as ***general homogeneous*** and ***particular inhomogeneous*** solutions. The homogenous solution is known up to some constants which need to be determined. The particular solution is any function $x_p(t)$ which satisfies the equation of motion, i.e., $\ddot{x}_p(t) + 2\,\zeta\,\omega\,\dot{x}_p(t) + \omega^2\,x_p(t) = f(t)/m$ is satisfied at all times. This solution can be found simply in some cases but has to be constructed in most realistic situations.

The formulation of the equations of motion for SDOF systems and the solutions of these equations require to select systems of coordinates, account for solution properties, and satisfy initial conditions. The following items provide details on these topics.

1. The equations of motions are meaningless in the absence of system of coordinates. Moreover, the functional forms of these equations and their solutions depend on the system of coordinates. For example, in the y-coordinate defined by $y(t) = x(t) + a$, the equations of motion Eqs. 2.7 and 2.8 of the SDOF system in Fig. 2.2 take the form

$$m\,\ddot{y}(t) = f(t) - k\,\big(y(t) - a\big) - c\,\dot{y}(t) \tag{2.10}$$

   or, equivalently,

$$\ddot{y}(t) + 2\,\zeta\,\omega\,\dot{y}(t) + \omega^2\,x(t) = \big(f(t) + k\,a\big)/m \tag{2.11}$$

   with the initial conditions $y(0) = a + x_0$ and $\dot{y}(0) = \dot{x}_0$. Note that, if $f(t) = 0$, the SDOF system is in free vibration in the x-coordinate and forced vibration in the y-coordinate.
2. The functional forms of the general solutions of damped ($\zeta > 0$) and undamped ($\zeta = 0$) SDOF systems have similarities and notable differences.

   If $\zeta = 0$, the general homogeneous solution has the expression $x_h(t) = A\,\cos(\omega\,t) + B\,\sin(\omega\,t)$. It is periodic with period $T = 2\,\pi/\omega$ and oscillates indefinitely. The displacement function of these systems is

$$x(t) = A\,\cos(\omega\,t) + B\,\sin\cos(\omega\,t) + x_p(t), \quad t \geq 0. \tag{2.12}$$

Note that undamped systems are not realistic since all mechanical/structural systems dissipate energy during deformation so that damping is always present, although it can be very small.

If $\zeta \in (0, 1)$, the displacement has the form

$$x(t) = x_h(t) + x_p(t) = e^{-\zeta\,\omega\,t}\Big(A\,\cos(\omega_d\,t) + B\,\sin(\omega_d\,t)\Big) + x_p(t), \quad t \geq 0, \tag{2.13}$$

where $\omega_d = \omega\sqrt{1-\zeta^2}$. The bracket in the expression of the homogeneous solution is periodic with period $T_d = 2\pi/\omega_d$. However, the homogeneous solution is not periodic. It decays to zero as time increases indefinitely so that $x(t) \simeq x_p(t)$ for large times. This limit case is referred to as the ***steady-state solution***. Similar considerations hold for damping ratios $\zeta \geq 1$ (see Sect. 2.4.1).

3. The ICs have to be imposed on the system solution $x(t) = x_h(t) + x_p(t)$, i.e., $x(0) = x_h(0) + x_p(0) = x_0$ and $\dot{x}(0) = \dot{x}_h(0) + \dot{x}_p(0) = \dot{x}_0$. These conditions deliver the undetermined constants in the expression of the general solution $x(t)$.

   For example, consider a SDOF system subjected to a constant action, i.e., $f(t)$ is a constant q so that Eq. 2.8 becomes

$$\ddot{x}(t) + 2\,\zeta\,\omega\,\dot{x}(t) + \omega^2\,x(t) = q/m, \quad t \geq 0. \tag{2.14}$$

This equation suggests that we can assume that $x_p(t) = c$ is a constant. Indeed, this trial particular solution and Eq. 2.14 give $\omega^2\,c = q/m$ so that $c = q/\big(m\,\omega^2\big) = q/k = x_{\text{st}}$ is the deformation of the system under the static load q. The general solution is

$$x(t) = e^{-\zeta\,\omega\,t}\big[A\,\cos(\omega_d\,t) + B\,\sin(\omega_d\,t)\big] + x_{\text{st}}, \quad t \geq 0, \tag{2.15}$$

so that

$$x(t) = e^{-\zeta\,\omega\,t}\left[(x_0 - x_{\text{st}})\,\cos(\omega_d\,t) + \frac{\dot{x}_0 + \zeta\,\omega\,(x_0 - x_{\text{st}})}{\omega_d}\,\sin(\omega_d\,t)\right] + x_{\text{st}}, \quad t \geq 0. \tag{2.16}$$

by imposing the initial conditions $x(0) = x_0$ and $\dot{x}(0) = \dot{x}_0$ which require $A + x_{\text{st}} = x_0$ and $-\zeta\,\omega\,A + B\,\omega_d = \dot{x}_0$ or $A = x_0 - x_{\text{st}}$ and $B = \big(\dot{x}_0 + \zeta\,\omega\,(x_0 - x_{\text{st}})\big)/\omega_d$. The steady-state solution $x_{\text{ss}}(t) = x_{\text{st}}$ since $x(t) \to x_{\text{st}}$ as $t \to \infty$. The rate of the convergence to x_{st} is controlled by the product $\zeta\,\omega$.

Example 2.2 The free vibration solution of the oscillator in Eq. 2.8 and the initial conditions $(x_0, \dot{x}_0)$ result from Eq. 2.16 by setting $x_{\text{st}} = 0$ and have the expression

$$x_h(t) = e^{-\zeta\,\omega\,t}\left[x_0\,\cos(\omega_d\,t) + \frac{\dot{x}_0 + \zeta\,\omega\,x_0}{\omega_d}\,\sin(\omega_d\,t)\right].$$

In the y-system of coordinates, the solution satisfies an inhomogeneous equation which consists of the sum of the general solution of the homogeneous equation $y_h(t)$ and a particular solution of the inhomogeneous equation $y_p(t)$, i.e.,

$$y(t) = y_h(t) + y_p(t) = e^{-\zeta\,\omega\,t}\left[A\,\cos(\omega_d\,t) + B\,\sin(\omega_d\,t)\right] + a$$

since $y_p(t) = a$ satisfies the equation of motion so that it is a particular solution. The initial conditions $y(0) = x_0 + a$ and $\dot{y}(0) = \dot{x}_0$ give $x_0 + a = A + a$ and $-\zeta\,\omega\,A + B\,\omega_d = \dot{x}_0$ so that $A = x_0$, $B = \left(\dot{x}_0 + \zeta\,\omega\,x_0\right)/\omega_d$ and

$$y(t) = e^{-\zeta\,\omega\,t}\left[x_0\,\cos(\omega_d\,t) + \frac{\dot{x}_0 + \zeta\,\omega\,x_0}{\omega_d}\,\sin(\omega_d\,t)\right] + a$$

which is consistent with the free vibration solution $x(t)$ since $y(t) = x(t) + a$.

This simple example illustrates clearly that equations of motions are meaningless if system of coordinates is not specified and that the nature of the solution depends on the system of coordinates. In this example, the oscillator is in free and forced vibration in the x-coordinate and the y-coordinate.

The following two subsections construct general homogeneous solutions for damping ratios $\zeta > 1$, $\zeta = 1$, and $\zeta < 1$ (Sect. 2.4.1) and provide details on these solutions for damping ratios $\zeta < 1$ (Sect. 2.4.2). Sections 2.4.3 and 2.4.4 construct particular inhomogeneous solutions for simple and arbitrary forcing functions. The construction of particular solutions for arbitrary forcing functions uses the Duhamel integral. The responses of undamped and damped SDOF systems to harmonic forces are examined in Sects. 2.4.5 and 2.4.6.

2.4.1 General Homogeneous Solution

The general form of the homogeneous solution is $\exp\left(\lambda\,t\right)$. The parameter λ is determined by requiring that this function satisfies the homogeneous version of Eq. 2.8, which gives (see Appendix B)

$$\lambda^2\,e^{\lambda\,t} + 2\,\zeta\,\omega\,\lambda\,e^{\lambda\,t} + \omega^2\,e^{\lambda\,t} = 0 \quad \text{or} \quad \left(\lambda^2 + 2\,\zeta\,\omega\,\lambda + \omega^2\right)e^{\lambda\,t} = 0, \quad t \geq 0.$$

Since the equality has to hold at all times and the exponential $\exp\left(\lambda\,t\right)$ is not zero in finite times for bounded λ, the above condition is satisfied only if $\lambda^2 + 2\,\zeta\,\omega\,\lambda + \omega^2 = 0$. This is second degree polynomial in λ whose roots can be real or complex. The general solution of the homogenous equation has the form

$$x_h(t) = \begin{cases} c_1\,e^{\lambda_1\,t} + c_2\,e^{\lambda_2\,t}, & \text{if } \lambda_1 \neq \lambda_2 \\ c_1\,e^{\lambda_1\,t} + c_2\,t\,e^{\lambda_1\,t}, & \text{if } \lambda_1 = \lambda_2, \end{cases} \tag{2.17}$$

where λ_1 and λ_2 are the roots of $\lambda^2 + 2\,\zeta\,\omega\,\lambda + \omega^2 = 0$. These roots are

$$\lambda_{1,2} = -\zeta\,\omega \pm \sqrt{(\zeta\,\omega)^2 - \omega^2} = \begin{cases} -\zeta\,\omega \pm \omega\sqrt{\zeta^2 - 1}, & \text{if } \zeta > 1 \\ -\omega, & \text{if } \zeta = 1 \\ -\zeta\,\omega \pm i\,\omega\sqrt{1 - \zeta^2}, & \text{if } \zeta < 1, \end{cases} \tag{2.18}$$

where $i = \sqrt{-1}$ is the imaginary unit. Systems with $\zeta > 1$, $\zeta = 1$, and $\zeta < 1$ are said to be over-damped, critically-damped, and under-damped. The general homogeneous solutions of these systems are

$$x_h(t) = \begin{cases} c_1\,\exp\left((-\zeta\,\omega + \omega\sqrt{\zeta^2 - 1})\,t\right) + c_2\,\exp\left((-\zeta\,\omega - \omega\sqrt{\zeta^2 - 1})\,t\right), & \text{if } \zeta > 1 \\ c_1\,\exp\left(-\,\omega\,t\right) + c_2\,t\,\exp\left(-\omega\,t\right), & \text{if } \zeta = 1 \\ c_1\,\exp\left((-\zeta\,\omega + i\,\omega\sqrt{1 - \zeta^2})\,t\right) + c_2\,\exp\left((-\zeta\,\omega - i\,\omega\sqrt{1 - \zeta^2})\,t\right), & \text{if } \zeta < 1. \end{cases} \tag{2.19}$$

The form of the homogeneous solution depends strongly on the damping parameter ζ which was previously introduced as just a notation. The above solutions show that the notation ζ has a deep physical meaning. It is common to refer to ζ as ***damping ratio***.

- ***Case 1.*** $\boldsymbol{\zeta > 1}$: Since $\sqrt{\zeta^2 - 1} < \zeta$, the exponents $\lambda_{1,2} = -\zeta\,\omega \pm \omega\sqrt{\zeta^2 - 1}$ are negative so that $x_h(t) \to 0$ as $t \to \infty$. The homogenous solution of over-damped SDOF has no oscillations. It decreases in time at rates depending on ζ and ω. The constants in the expression of $x_h(t)$ result from the ICs $(x_0, \dot{x}_0)$, which give $x_h(0) = x_0 = c_1 + c_2$ and $\dot{x}_h(0) = \dot{x}_0 = c_1\,\lambda_1 + c_2\,\lambda_2$.
- ***Case 2.*** $\boldsymbol{\zeta = 1}$: The homogeneous solution of critically-damped SDOF systems has a similar behavior, i.e., $x_h(t) \to 0$ as $t \to \infty$ and exhibits no oscillations. The constants in the expression of $x_h(t)$ result from the ICs $(x_0, \dot{x}_0)$, which give $x_h(0) = x_0 = c_1$ and $\dot{x}_h(0) = \dot{x}_0 = -c_1\,\omega + c_2$.
- ***Case 3.*** $\boldsymbol{\zeta < 1}$: An alternative form of the homogeneous solution is

$$\begin{aligned} x_h(t) &= e^{-\zeta\,\omega\,t}\left(c_1\,e^{i\,\omega_d\,t} + c_2\,e^{-i\,\omega_d\,t}\right) \\ &= e^{-\zeta\,\omega\,t}\left(A\,\cos(\omega_d\,t) + B\,\sin(\omega_d\,t)\right), \end{aligned} \tag{2.20}$$

where $\omega_d = \omega\sqrt{1 - \zeta^2}$. The latter expression follows from the identity $\exp(\pm i\,u) = \cos(u) \pm i\,\sin(u)$ which holds for any real u. Note that the constants (c_1, c_2) and (A, B) are complex- and real-valued.

The homogeneous solution in this case differs significantly from those of the previous two cases. In contrast two homogeneous solutions for cases 1 and 2 which exhibit no oscillations, $x_h(t)$ in Eq. 2.20 has oscillations of frequency ω_d with amplitudes decreasing in time at the rate $\zeta\,\omega$. Case 3 is particularly relevant for applications since ζ is much smaller than unity for almost all structural/mechanical systems.

We conclude with the observations that (1) the general homogeneous solution $x_h(t)$ has the physical meaning of free vibration and (2) the features of the free vibration solution are determined by damping, i.e., $x_h(t) \to 0$ as time $t \to \infty$ if $\zeta > 0$, $x_h(t)$ has no oscillation if $\zeta \geq 1$, $x_h(t)$ exhibits oscillation if $\zeta < 1$, and $x_h(t)$ is periodic with period $2\pi/\omega$ if $\zeta = 0$ (see Eq. 2.20). The following subsection examines in details under-damped SDOF systems.

2.4.2 General Homogeneous Solution ($\zeta < 1$)

The general solution of Eq. 2.8 with $f(t) = 0$ and $\zeta < 1$ constitutes the free vibration solution of an under-damped SDOF system and has the expression $x_h(t)$ in Eq. 2.20. These types of systems are commonly encountered in applications. The section presents the standard form of the general homogeneous solution and its amplitude-phase version. It also illustrates a method for estimating the damping ratio ζ from measurements.

2.4.2.1 Standard Form

The constants A and B in the expression of $x_h(t)$ can be found by imposing the initial conditions $x_h(0) = x_0$ and $\dot{x}_h(0) = \dot{x}_0$. These conditions give $A = x_0$ from the expression of $x_h(t)$ and $-\zeta\,\omega\,A + B\,\omega_d = \dot{x}_0$ from $\dot{x}_h(t) = -\zeta\,\omega\,e^{-\zeta\,\omega t}\big(A\,\cos(\omega_d\,t) + B\,\sin(\omega_d\,t)\big) + e^{-\zeta\,\omega t}\big(-A\,\omega_d\,\sin(\omega_d\,t) + B\,\omega_d\,\cos(\omega_d\,t)\big)$. This system of linear equations gives $A = x_0$ and

$$B = \frac{\dot{x}_0 + \zeta\,\omega\,A}{\omega_d} = \frac{\dot{x}_0 + \zeta\,\omega\,x_0}{\omega_d}$$

so that

$$x_h(t) = x_{\text{free vibration}}(t) = e^{-\zeta\,\omega t}\left[x_0\,\cos(\omega_d\,t) + \frac{\dot{x}_0 + \zeta\,\omega\,x_0}{\omega_d}\,\sin(\omega_d\,t)\right]. \tag{2.21}$$

This is the most general expression of the free vibration solution. It can be used to find the free vibration solutions in special cases. For example, the free vibration solution of undamped systems ($c = 0$ which implies $\zeta = 0$) is

$$x_h(t) = x_{\text{free vibration}}(t) = x_0\,\cos(\omega\,t) + \frac{\dot{x}_0}{\omega}\,\sin(\omega\,t). \tag{2.22}$$

There is a significant difference between the free vibration solutions of Eqs. 2.21 and 2.22. The free vibration in Eq. 2.22 is periodic with period $T = 2\pi/\omega$ while

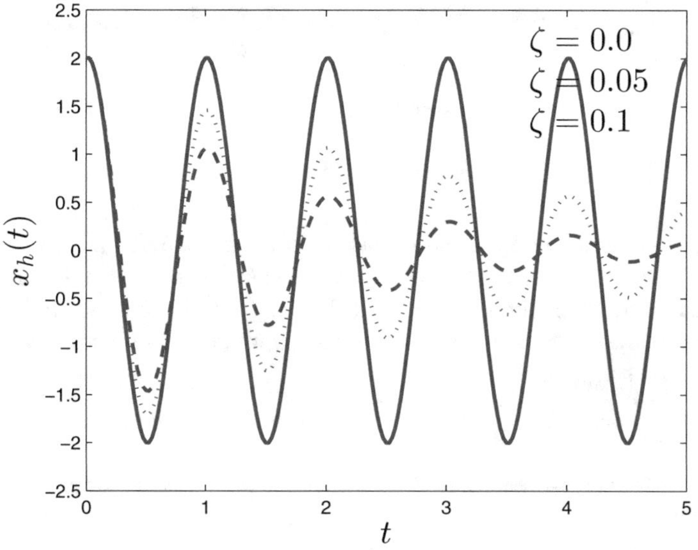

Fig. 2.3 Free vibration solution for $\zeta = 0.0$, 0.05, and 0.1

that of Eq. 2.21 is not since the motion amplitude decreases steadily in time. The parameters T and ω are referred to as the ***natural period*** and the ***natural frequency*** of the oscillator.

Example 2.3 The solid, dotted, and dashed lines of Fig. 2.3 show the free vibration solutions of SDOF systems with natural frequencies $\omega = 2\pi$ and damping ratios $\zeta = 0.0$, 0.05, and 0.1 for the initial conditions $x_0 = 2$ and $\dot{x}_0 = 1$. All solutions start at $x_0 = 2$ and initially have positive slope although it is not highly visible at the figure scale. The amplitudes decrease at rates which increase with the damping ratio. The motion is periodic with period $T = 1$ for $\zeta = 0$ but not for $\zeta > 0$.

2.4.2.2 Amplitude-Phase Representation

An alternative form of $x_h(t)$ for $\zeta < 1$ is

$$x_h(t) = a\, e^{-\zeta\,\omega\,t}\, \sin\left(\omega_d\, t + \varphi\right), \tag{2.23}$$

where the amplitude a and the phase φ have the expressions

$$a^2 = x_0^2 + \left(\frac{\dot{x}_0 + \zeta\,\omega\, x_0}{\omega_d}\right)^2 \quad \text{and} \quad \tan(\varphi) = \frac{\dot{x}_0 + \zeta\,\omega_d\, x_0}{\omega\, x_0}. \tag{2.24}$$

This follows from the observation that $x_h(t)$ of Eq. 2.21 can be written as

$$
\begin{aligned}
x_h(t) &= e^{-\zeta\,\omega\,t}\,a\left(\frac{x_0}{a}\,\cos(\omega_d\,t) + \frac{\dot{x}_0 + \zeta\,\omega\,x_0}{\omega_d\,a}\,\sin(\omega_d\,t)\right)\\
&= a\,e^{-\zeta\,\omega\,t}\left(\sin(\varphi)\,\cos(\omega_d\,t) + \cos(\varphi)\,\sin(\omega_d\,t)\right)\\
&= a\,e^{-\zeta\,\omega\,t}\,\sin\big(\omega_d\,t + \varphi\big).
\end{aligned}
$$

The notations $\sin(\varphi) = x_0/a$ and $\cos(\varphi) = \big(\dot{x}_0 + \zeta\,\omega\,x_0\big)/\big(\omega_d\,a\big)$ are meaningful since the absolute values of these ratios are smaller than unity and the sum of their squares is unity. The latter equality results by using a trigonometric identity.

We summarize some useful features of the amplitude-phase representation of the free vibration solution.

1. The largest value of $x_h(t)$ and its derivatives are readily available. They are a, $a\,\omega_d$, and $a\,\omega_d^2$ for $x_h(t)$, $\dot{x}_h(t)$, and $\ddot{x}_h(t)$.
2. The graphical representation of $x_h(t)$, $\dot{x}_h(t)$, and $\ddot{x}_h(t)$ is simple as they are modulated periodic functions. For example, $x_h(t)$ is a sine wave shifted by the phase φ which is modulated by the exponential amplitude $a\,\exp\big(-\zeta\,\omega\,t\big)$.
3. For undamped systems, $x_h(t) = a\,\sin(\omega\,t + \varphi)$ so that the elastic and kinetic energies are $\mathrm{SE}(t) = (1/2)\,k\,x_h(t)^2 = (1/2)\,k\,a^2\,\sin^2(\omega\,t + \varphi)$ and $\mathrm{KE}(t) = (1/2)\,m\,\dot{x}_h(t)^2 = (1/2)\,m\,a^2\,\omega^2\,\cos^2(\omega\,t + \varphi)$ so that, since $m\,\omega^2 = k$, we have $\mathrm{SE}(t) + \mathrm{KE}(t) = k\,a^2/2 =$ constant, as expected since the system is conservative.
4. The velocity $\dot{x}_h(t)$ and acceleration $\ddot{x}_h(t)$ are out of phase by $\pi/2$ and π relative to $x_h(t)$. These relationships follow from properties of trigonometric functions. We have

$$
\begin{aligned}
\dot{x}_h(t) &= a\,\omega\,\cos(\omega\,t + \varphi) = a\,\omega\,\sin(\pi/2 - \omega\,t - \varphi)\\
&= -a\,\omega\,\sin(\omega\,t + \varphi - \pi/2)\\
&= a\,\omega\,\sin(\omega\,t + \varphi + \pi/2)
\end{aligned}
$$

and

$$
\ddot{x}_h(t) = -a\,\omega^2\,\sin(\omega\,t + \varphi) = a\,\omega^2\,\sin(\omega\,t + \varphi + \pi)
$$

so that $\dot{x}_h(t)$ and $\ddot{x}_h(t)$ are out of phase relative to $x_h(t)$ by $\pi/2$ and π. Their amplitudes are $a\,\omega$ and $a\,\omega^2$. The solid, dashed, and dotted lines of Fig. 2.4 show the solutions $x_h(t)$, $\dot{x}_h(t)$, and $\ddot{x}_h(t)$ over a few periods for $a = 1$, $\omega = \pi$, and $\varphi = \pi/4$. The plots illustrate the phase difference between these three functions. The initial values of these solutions are $x_h(0) = \sin(\pi/4) = 0.7071$, $\dot{x}_h(0)/\omega = \sin(\pi/4 + \pi/2) = 0.7071$, and $\ddot{x}_h(0) = \sin(\pi/4 + \pi) = -0.7071$.

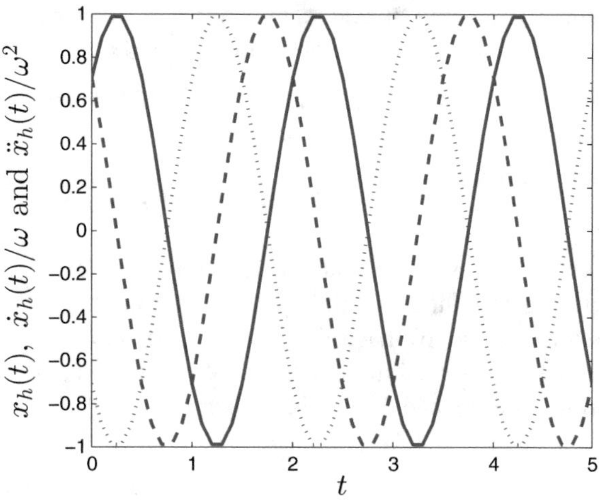

Fig. 2.4 Solutions $x_h(t)$, $\dot{x}_h(t)$ and $\ddot{x}_h(t)$ for $\omega = \pi$ and $\varphi = \pi/4$ (solid, dashed, and dotted lines)

2.4.2.3 Damping Estimation

The free vibration displacement of a damped SDOF system is the product of a periodic function with period $T_d = 2\pi/\omega_d$, the square bracket in Eq. 2.21, and the exponential function $\exp(-\zeta\,\omega\,t)$. Since the square bracket has the same values at the times t and $t + k\,T_d$, where k is an integer, we have

$$\frac{x(t)}{x(t + k\,T_d)} = \frac{e^{-\zeta\,\omega\,t}}{e^{-\zeta\,\omega\,(t+k\,T_d)}} = e^{\zeta\,\omega\,(k\,T_d)}$$

so that

$$\delta = \ln\left(\frac{x(t)}{x(t + k\,T_d)}\right) = \zeta\,\omega\,(k\,T_d) = \frac{2\pi\,k\,\zeta}{\sqrt{1-\zeta^2}} \simeq 2\pi\,k\,\zeta, \qquad (2.25)$$

where the latter approximations hold for $\zeta \ll 1$. These relationships can be used to estimate the damping ratio ζ from measurements of the displacement at time intervals equal to T_d or multiple of it. The measurements are used to estimate δ. The corresponding value of ζ results from Eq. 2.25.

2.4.3 *Particular Inhomogeneous Solution, Simple Forcing Functions*

We first construct particular solutions for several simple forcing functions and then use the resulting solutions to develop a method for finding particular solution for arbitrary forcing functions. In contrast to standard methods which are abstract (see Appendix B), our approach is intuitive and based on mechanics.

Example 2.4 Consider an undamped SDOF which is subjected to a constant forcing function $f(t) = q, t \geq 0$, and has the ICs $(x_0, \dot{x}_0)$. The general solution has the form

$$x(t) = A\,\cos(\omega\,t) + B\,\sin(\omega\,t) + x_p(t), \quad t \geq 0,$$

where the particular solution is $x_p(t) = q/\left(m\,\omega^2\right) = q/k = x_{\text{st}}$. That this is a particular solution results by checking that it satisfies the equation of motion $\ddot{x} + \omega^2\,x = q/m$. We have $\ddot{x}_p + \omega^2\,x_p = 0 + \omega^2\,(q/k) = q/m$ so that $x_p(t) = q/k$ is a particular solution. The initial conditions $x(0) = x_0$ and $\dot{x}(0) = \dot{x}_0$ imply $A + x_{\text{st}} = x_0$ and $B\,\omega = \dot{x}_0$ so that

$$x(t) = \left(x_0 - x_{\text{ st}}\right)\,\cos(\omega\,t) + \frac{\dot{x}_0}{\omega}\,\sin(\omega\,t) + x_{\text{ st}}.$$

For zero initial conditions, the displacement $x(t) = x_{\text{ st}}\left(1 - \cos(\omega\,t)\right)$ is shown in Fig. 2.5 for $\omega = \pi$. Note that the dynamic amplification factor DAF $= \max_t\{x(t)\}/x_{\text{st}} = 2$. This means that the response of system to a suddenly applied force q is twice that of the system if the force is applied statically.

Example 2.5 Consider a damped SDOF which is at rest at the initial time $t = 0$, i.e., $x_0 = 0$ and $\dot{x}_0 = 0$, and is subjected to a forcing function which is constant and equal to q in a time interval (t_1, t_2) and zero outside this interval (see Fig. 2.6, left panel).

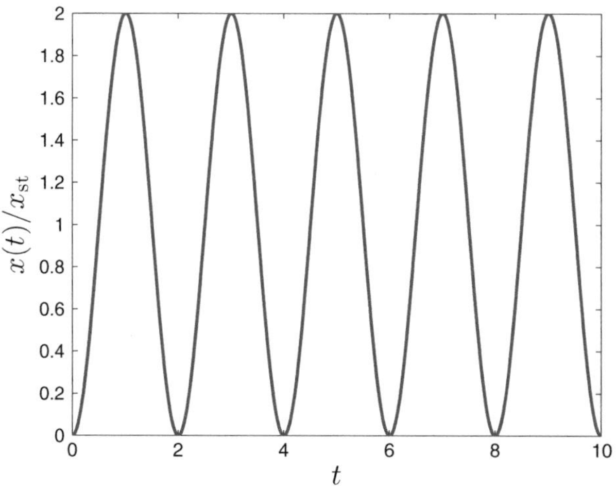

Fig. 2.5 Displacement of an undamped SDOF system under a suddenly applied force

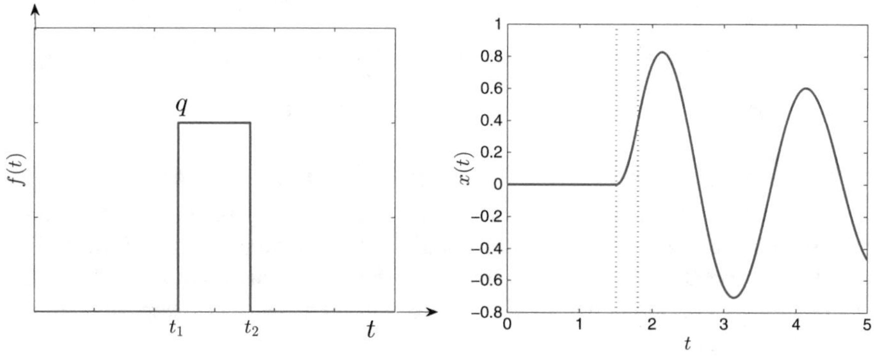

Fig. 2.6 Forcing function $f(t) = q$ in (t_1, t_2) and zero outside this time interval (left panel) and displacement $x(t)$ for $x_{st} = 1$, $\omega = \pi$, $\zeta = 0.05$, $t_1 = 1.5$, and $t_2 = 1.8$ (right panel)

The oscillator remains at rest till t_1 since it has zero ICs and is subjected to no action. It exhibits forced and free vibrations for times t in (t_1, t_2) and times $t \geq t_2$ larger than t_2. The general solution in (t_1, t_2) is the sum of the general homogeneous solution and a particular solution, i.e.,

$$x(s) = e^{-\zeta\,\omega\, s}\left(A\,\cos(\omega_d\, s) + B\,\sin(\omega_d\, s)\right) + x_p(s), \quad 0 < s < t_2 - t_1,$$

where $s = t - t_1 > 0$ denotes a local time which is used for convenience and $x_p(s) = x_{st} = q/k$ from Example 2.4. The ICs $x(0) = 0$ and $\dot{x}(0) = 0$ imply

$$x(0) = 0 \Longrightarrow A + x_{st} = 0 \quad \text{and}$$
$$\dot{x}(0) = 0 \Longrightarrow -\zeta\,\omega\, A + B\,\omega_d = 0$$

so that $A = -x_{st}$, $B = -\zeta\,\omega\, x_{\,st}/\omega_d = -\zeta\, q/\left(k\sqrt{1-\zeta^2}\right)$, and

$$x(s) = x_{\,st}\left[1 - e^{-\zeta\,\omega\, s}\left(\cos(\omega_d\, s) + \frac{\zeta}{\sqrt{1-\zeta^2}}\,\sin(\omega_d\, s)\right)\right], \quad 0 < s < t_2 - t_1. \tag{2.26}$$

The oscillator displacement and velocity at the end of the time interval (t_1, t_2), i.e., $x(\Delta t)$ and $\dot{x}(\Delta t)$, result from the above equation with $s = \Delta t = t_2 - t_1$. The free vibration solution which occurs for times $t \geq t_2$ results from Eq. 2.21 and has the form

$$x(t) = e^{-\zeta\,\omega\, u}\left[x(\Delta t)\,\cos(\omega_d\, u) + \frac{\dot{x}(\Delta t) + \zeta\,\omega\, x(\Delta t)}{\omega_d}\,\sin(\omega_d\, u)\right], \tag{2.27}$$

where $u = t - t_2 \geq 0$ denotes the time measured from the end of the time interval (t_1, t_2). In summary, the oscillator displacement has the expression

$$x(t)=\begin{cases} 0, & \text{for } t\leq t_1 \\ x\,\mathrm{st}\left[1-e^{-\zeta\,\omega\,(t-t_1)}\left(\cos(\omega_d\,(t-t_1))+\frac{\zeta}{\sqrt{1-\zeta^2}}\sin(\omega_d\,(t-t_1))\right)\right], & \text{for } t_1<t<t_2 \\ e^{-\zeta\,\omega\,(t-t_2)}\left[x(\Delta t)\,\cos(\omega_d\,(t-t_2))+\frac{\dot{x}(\Delta t)+\zeta\,\omega\,x(\Delta t)}{\omega_d}\sin(\omega_d\,(t-t_2))\right], & \text{for } t\geq t_2. \end{cases} \tag{2.28}$$

The right panel of Fig. 2.6 shows the oscillator displacement in Eq. 2.28 for $x_{st} = 1$, $\omega = \pi$, $\zeta = 0.05$, $t_1 = 1.5$, and $t_2 = 1.8$. There is no response prior to t_1 since the oscillator is at rest at the initial time and there is no force acting on it till t_1. The oscillator responds to the force $f(t) = q$ in the time interval (t_1, t_2) and vibrates freely after t_2. The rate of decrease of the free vibration solution is controlled by the damping ratio ζ.

2.4.4 *Particular Inhomogeneous Solution, Arbitrary Forcing Functions*

In contrast to developments of the previous section which construct particular solutions for special forcing functions, the Duhamel integral delivers these solutions for arbitrary forcing functions. This section constructs this integral by using results of Example 2.5 and the linearity of the equation of motion according to which a system response to a set of inputs is given by the sum of responses to the individual inputs in this set.

Suppose that the duration $\Delta t = t_2 - t_1$ of the forcing function in Example 2.5 is such that $\omega\,\Delta t,\ \omega_d\,\Delta t \ll 1$. Under this condition, we have (see Appendix A on Taylor's formula)

$$e^{-\zeta\,\omega\,\Delta t} \simeq 1 - \zeta\,\omega\,\Delta t + O\big(\omega\,\Delta t\big)^2$$
$$\cos\big(\omega_d\,\Delta t\big) \simeq 1 + O\big(\omega_d\,\Delta t\big)^2$$
$$\sin\big(\omega_d\,\Delta t\big) \simeq \omega_d\,\Delta t + O\big(\omega_d\,\Delta t\big)^3,$$

where $O\big(\omega\,\Delta t\big)^r$ denotes terms proportional to $(\omega\,\Delta t)^k$, $k \geq r$, which are very small under the assumption $\omega\,\Delta t \ll 1$. Note that the terms $\omega\,\Delta t$ and $\omega_d\,\Delta t$ have the same order of magnitude since $\omega_q = \omega\sqrt{1-\zeta^2}$ and ζ is small. The above approximations and the expression of $x(s)$ in Eq. 2.26 give $x(\Delta t) \sim O\big(\omega\,\Delta t\big)^2$ and $\dot{x}(\Delta t) \sim (q\,\Delta t)/m + O\big(\omega\,\Delta t\big)^2$. For example, the square bracket in Eq. 2.26 is

$$1-\left(1-\zeta\,\omega\,\Delta t+O\big(\omega\,\Delta t\big)^2\right)$$
$$\times\left(1+O\big(\omega_d\,\Delta t\big)^2+\frac{\zeta}{\sqrt{1-\zeta^2}}\left(\omega_d\,\Delta t+O\big(\omega_d\,\Delta t\big)^3\right)\right)$$

$$\sim 1-\left(1-\zeta\,\omega\,\Delta t+O\big(\omega\,\Delta t\big)^2\right)\left(1+\zeta\,\omega\,\Delta t+O\big(\omega_d\,\Delta t\big)^2\right)\sim O\big(\omega_d\,\Delta t\big)^2.$$

The displacement and velocity $x(\Delta t)\sim O\big(\omega\,\Delta t\big)^2$ and $\dot{x}(\Delta t)\sim(q\,\Delta t)/m+O\big(\omega\,\Delta t\big)^2$ provide the input for the free vibration solution for $t>t_2$ under the assumption $\omega\,\Delta t,\ \omega_d\,\Delta t\ll 1$. The reader is encouraged to construct these approximations by using the above approximations of $\exp\big(-\zeta\,\omega\,\Delta t\big)$, $\cos\big(\omega_d\,\Delta t\big)$, and $\sin\big(\omega_d\,\Delta t\big)$.

The latter result has an important physical meaning. It represents the displacement of an oscillator following an impulse of duration Δt and intensity q. This interpretation provides the ingredient for constructing particular solutions for SDOF systems subjected to arbitrary forcing functions $f(t)$. The construction involves the following three steps.

- *Step 1:* Discretize the support of the forcing function $f(t)$ in small time intervals Δu such that $\omega\,\Delta u,\ \omega_d\,\Delta u\ll 1$ and represent $f(t)$ by a piecewise constant function equal to $f(u)$ in the time interval $[u,u+\Delta u)$, as illustrated in Fig. 2.7, i.e., the forcing function $f(t)$ for t in $[u,u+\Delta u)$ is approximated by its value $f(u)$ at the left end of this time interval.
- *Step 2:* The free vibration solution of the oscillator following the forcing function which is constant and equal to $f(u)$ in the time interval $(u,u+\Delta u)$ and zero otherwise results from the last expression in Eq. 2.28 and has the form

$$\begin{aligned}\Delta x(t;u)&\sim e^{-\zeta\,\omega\,(t-u-\Delta u)}\Big[O(\omega_d\,\Delta u)^2\cos\big(\omega_d\,(t-u-\Delta u)\big)\\&\quad+\frac{f(u)\,\Delta u/m+O(\omega_d\,\Delta u)^2}{\omega_d}\sin\big(\omega_d\,(t-u-\Delta u)\big)\Big]\\&\sim e^{-\zeta\,\omega\,(t-u-\Delta u)}\,\frac{f(u)\,\Delta u/m}{\omega_d}\sin\big(\omega_d\,(t-u-\Delta u)\big)+O(\omega_d\,\Delta u)^2\end{aligned}$$

  under our assumption of small Δu.
- *Step 3:* Add the contributions to the oscillator displacement caused by the forcing function prior to the time t, i.e.,

$$x(t)\simeq\sum_{u+\Delta u\le t}\Delta x(t;u)=\sum_{u+\Delta u\le t}e^{-\zeta\,\omega\,(t-u-\Delta u)}\,\frac{f(u)}{m\,\omega_d}\sin\big(\omega_d\,(t-u-\Delta u)\big)\,\Delta u,$$

  and take the limit as $\Delta u\to 0$, which gives

$$x(t)=\frac{1}{m\,\omega_d}\int_0^t e^{-\zeta\,\omega\,(t-u)}\sin\big(\omega_d\,(t-u)\big)\,f(u)\,du. \tag{2.29}$$

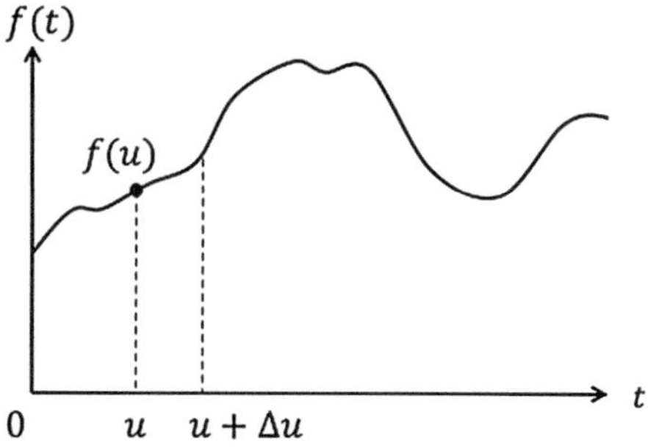

Fig. 2.7 Discretization of an arbitrary forcing function $f(t)$

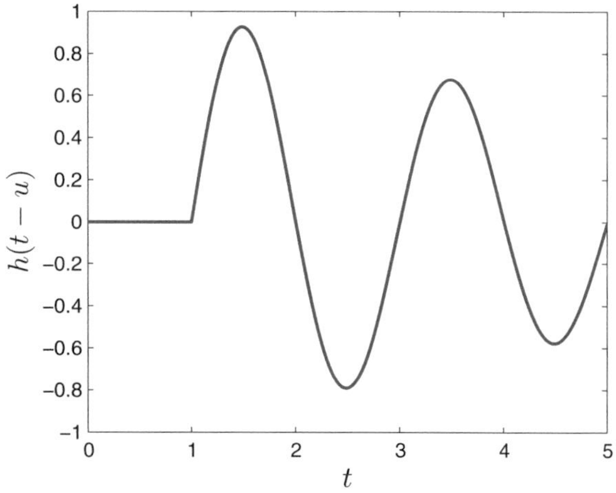

Fig. 2.8 Unit impulse response function $h(t-u)$ for $u=1, m=1, \omega=\pi$, and $\zeta=0.05$

We conclude with the following comments. First, it is common to write the above formula as

$$x(t) = \int_0^t h(t-u)\, f(u)\, du, \qquad \text{where}$$
$$h(t-u) = \frac{1}{m\,\omega_d}\, e^{-\zeta\,\omega\,(t-u)} \sin\big(\omega_d\,(t-u)\big), \tag{2.30}$$

and refer to it as the ***Duhamel integral***. The kernel $h(t-u)$, $t \geq u$, referred to as the ***unit impulse response function***, is the free vibration solution of the oscillator at time $t \geq u$ caused by a unit impulse at time u. The unit impulse response function is shown in Fig. 2.8 for a SDOF system with $u = 1$, $m = 1$, $\omega = \pi$, and $\zeta = 0.05$. Note that the unit impulse response function constitutes the limit of the free vibration solution shown in Fig. 2.6 following the impulse in the time interval (t_1, t_2) as its duration decreases to zero, i.e., $t_2 - t_1 \to 0$.

Second, $x(t)$ in Eq. 2.30 is a particular solution since it satisfies the inhomogeneous equation of motion of the oscillator. Accordingly, we use the notation $x_p(t) = x(t)$. It is not the general solution since it does not include the free vibration component. Note also that it cannot satisfy non-zero initial conditions. For example, we have $x(0) = 0$ from Eq. 2.30.

Third, the displacement of a SDOF subjected to an arbitrary forcing function $f(t)$ has the form

$$x(t) = e^{-\zeta\,\omega t}\left(A\,\cos(\omega_d\,t) + B\,\sin(\omega_d\,t)\right) + \int_0^t h(t-u)\,f(u)\,du, \qquad (2.31)$$

where the constants A and B can be determined from the initial conditions $x(0) = x_0$ and $\dot{x}(0) = \dot{x}_0$. Note also that the free vibration component of $x(t)$, i.e., $e^{-\zeta\,\omega t}\left(A\,\cos(\omega_d\,t) + B\,\sin(\omega_d\,t)\right)$, vanishes in time so that

$$x(t) \simeq x_{ss}(t) = \int_0^t h(t-u)\,f(u)\,du \qquad (2.32)$$

for large times. The solution $x_{ss}(t)$, referred to as the ***steady-state*** solution, provides a satisfactory approximations of $x(t)$ for sufficiently large times t for which $\exp(-\zeta\,\omega t) \ll 1$.

Example 2.6 Consider the undamped oscillator of Example 2.4 with zero initial conditions and subjected to a suddenly applied force q. The unit impulse response function of Eq. 2.30 becomes $h(t-u) = \sin\left(\omega\,(t-u)\right)/\left(m\,\omega\right)$ so that the particular solution is

$$x_p(t) = \int_0^t \frac{q}{m\,\omega}\,\sin\left(\omega\,(t-u)\right)du = \frac{q}{m\,\omega}\,\frac{\cos\left(\omega\,(t-u)\right)}{\omega}\bigg|_0^t = x_{st}\left(1-\cos(\omega\,t)\right)$$

as by direct calculations.

Example 2.7 Consider a damped oscillator with natural frequency ω and damping ratio ζ which is subjected to a suddenly applied constant force $f(t) = q$. The oscillator is at rest at the initial time, i.e., $x_0 = 0$ and $\dot{x}_0 = 0$. The particular solution, i.e., the Duhamel integral in Eq. 2.30, has the form

$$\begin{aligned} x_p(t) &= \frac{q}{m\,\omega_d}\int_0^t e^{-\zeta\,\omega\,(t-u)}\,\sin\left(\omega_d\,(t-u)\right)du = \frac{q}{m\,\omega_d}\int_0^t e^{-\zeta\,\omega\,s}\,\sin\left(\omega_d\,s\right)ds \\ &= \frac{q}{m\,\omega_d}\left[I(-\zeta\,\omega,\omega_d;t) - I(-\zeta\,\omega,\omega_d;0)\right] \end{aligned}$$

by the change of variable $s = t - u$, where

$$I(c,b;u) = \int e^{c\,u}\,\sin(b\,u)\,du = \frac{e^{c\,u}}{c^2+b^2}\left(c\,\sin(b\,u) - b\,\cos(b\,u)\right).$$

The constants A and B in Eq. 2.31 are zero since $x(0) = A = x_0 = 0$ and

$$\begin{aligned}\dot{x}_p(t) &= -\zeta\,\omega\, e^{-\zeta\,\omega\, t}\big(A\,\cos(\omega_d\, t) + B\,\sin(\omega_d\, t)\big)\\ &\quad + \omega_d\, e^{-\zeta\,\omega\, t}\big(-A\,\sin(\omega_d\, t) + B\,\cos(\omega_d\, t)\big)\\ &\quad + \frac{d}{dt}\int_0^t h(t-u)\, f(u)\, du\end{aligned}$$

and (see Appendix B)

$$\frac{d}{dt}\int_0^t h(t-u)\, f(u)\, du = \int_0^t \frac{\partial h(t-u)}{\partial t}\, f(u)\, du + h(0)\, f(t)$$

so that $\dot{x}(0) = -A + \omega_d\, B = \dot{x}_0 = 0$.

Example 2.8 Suppose that the oscillator in the previous example is subjected to a constant force during the time interval $[0, \tau]$, i.e., $f(u) = q\, 1(u \le \tau),\ u \ge 0$, where the indictor function $1(A) = 1$ if A is true and zero otherwise. The particular solution by the Duhamel integral is

$$x_p(t) = \int_0^t h(t-u)\, q\, 1(u \le \tau)\, du = q\int_0^{\min\{t,\tau\}} h(t-u)\, du,$$

so that it is available analytically by using the integral $I(c, b; u)$ above.

Example 2.9 Suppose the oscillator of the previous example is used to model a building which is subjected to a seismic ground acceleration $a(t)$. The equation of motion given by Eq. 2.7 takes the form $m\big(\ddot{x}(t) + a(t)\big) = -k\, x(t) - c\,\dot{x}(t)$ since the acceleration of the mass has two components, one caused by the motion of the support and one caused by the system deformation. With the notations of Eq. 2.8 we have

$$\ddot{x}(t) + 2\,\zeta\,\omega\,\dot{x}(t) + \omega^2\, x(t) = -a(t), \tag{2.33}$$

so that the solution of Eq. 2.31 holds provided that $-a(t)$ is substituted for $f(t)/m$. We have

$$x(t) = e^{-\zeta\,\omega\, t}\big(A\,\cos(\omega_d\, t) + B\,\sin(\omega_d\, t)\big) - \int_0^t h_a(t-u)\, a(u)\, du, \tag{2.34}$$

where

$$h_a(t-u) = \frac{1}{\omega_d}\, e^{-\zeta\,\omega\,(t-u)}\,\sin\big(\omega_d\,(t-u)\big) \tag{2.35}$$

denotes the unit impulse response function for the input ground acceleration. For zero initial conditions, $A = 0$ since $x(0) = A$ and $B = 0$ since

$$\dot{x}(t) = B\,e^{-\zeta\,\omega t}\big[-\zeta\,\omega\,\sin(\omega_d\,t) + \omega_d\,\sin(\omega_d\,t)\big] - \frac{d}{dt}\int_0^t h_a(t-u)\,a(u)\,du$$
$$= B\,e^{-\zeta\,\omega t}\big[-\zeta\,\omega\,\sin(\omega_d\,t) + \omega_d\,\sin(\omega_d\,t)\big]$$
$$-\frac{d}{dt}\int_0^t \frac{\partial h_a(t-u)\,a(u)}{\partial t}\,du + h_a(0)\,a(t)$$

so that $\dot{x}(0) = B\,\omega_d$. The expression of the time derivative of the integral $I(t) = \int_0^t h_a(t-u)\,a(u)\,du$ results by taking the limit of $[I(t+\Delta t) - I(t)]/\Delta t$ as $\Delta t \to 0$, see the Leibniz integral rule, see Appendix B.

2.4.5 *Harmonic Force, Undamped System*

We determine particular solutions $x_p(t)$ of Eq. 2.8 with $\zeta = 0$, i.e., forced vibration solutions defined by

$$\ddot{x}_p(t) + \omega^2\,x_p(t) = f(t)/m, \quad t \geq 0. \tag{2.36}$$

Example 2.10 Suppose that the forcing function is the sine wave $f(t) = q\,\sin(\nu\,t)$, $\nu > 0$, $\nu \neq \omega$, and that the SDOF system has no damping ($c = 0$ or, equivalently, $\zeta = 0$). The particular solution is

$$x_p(t) = \frac{x_{\text{st}}}{1 - (\nu/\omega)^2}\,\sin(\nu\,t), \quad \nu \neq \omega, \tag{2.37}$$

where $x_{\text{st}} = q/k$. It can be established by using the trial functional form $x_p(t) = \alpha\,\sin(\nu\,t)$ and determine whether there is a constant α such that the equation of motion $\ddot{x}_p(t) + \omega^2\,x_p(t) = \big(q/m\big)\,\sin(\nu\,t)$ is satisfied at all times. This gives

$$-\alpha\,\nu^2\,\sin(\nu\,t) + \omega^2\,\alpha\,\sin(\nu\,t) = \big(q/m\big)\,\sin(\nu\,t)$$

or $\big(-\nu^2 + \omega^2\big)\,\alpha\,\sin(\nu\,t) = \big(q/m\big)\,\sin(\nu\,t)$. Since this equality must hold at all times, we have $\alpha\big(-\nu^2 + \omega^2\big) = q/m$ so that

$$\alpha = \frac{q/m}{-\nu^2 + \omega^2} = \frac{q/m}{\omega^2\,\big(1 - (\nu/\omega)^2\big)} = \frac{x_{\text{st}}}{1 - (\nu/\omega)^2} \tag{2.38}$$

has the required property and the trial function with this α is a particular solution.

The absolute value of the ratio α/x_{st}, i.e.,

$$\text{DAF} = \frac{1}{\big|1 - (\nu/\omega)^2\big|}, \tag{2.39}$$

is called ***dynamic amplification factor*** (DAF). It converges to 1 and 0 as $\nu/\omega \to 0$ and $\nu/\omega \to \infty$. Physically, there is no dynamic amplification for forcing functions with $\nu \ll \omega$ and there is virtually no response for $\nu \gg \omega$. The response becomes unbounded in time as $\nu/\omega \to 1$, a phenomenon referred to as ***resonance***.

2.4.5.1 Resonance Phenomenon

Consider the undamped oscillator in Example 2.10 under the harmonic force $f(t) = q\,\sin(\nu\,t)$ with initial conditions $(x_0, \dot{x}_0)$. For $\nu \neq \omega$, the oscillator displacement is

$$x(t) = A\,\cos(\omega\,t) + B\,\sin(\omega\,t) + \frac{x_{\,\mathrm{st}}}{1-(\nu/\omega)^2}\,\sin(\nu\,t)$$

with the notation in Eq. 2.37. The initial conditions imply $x_0 = A$ and $\dot{x}_0 = B\,\omega + x_{\,\mathrm{st}}\,\nu/\big(1-(\nu/\omega)^2\big)$ so that the general solution has the form

$$x(t) = x_0\,\cos(\omega\,t) + \frac{\dot{x}_0}{\omega}\,\sin(\omega\,t) + \frac{x_{\mathrm{st}}}{1-(\nu/\omega)^2}\left(\sin(\nu\,t) - \frac{\nu}{\omega}\,\sin(\omega\,t)\right) \qquad (2.40)$$

which simplifies to

$$x(t) = \frac{x_{\mathrm{st}}}{1-(\nu/\omega)^2}\left(\sin(\nu\,t) - \frac{\nu}{\omega}\,\sin(\omega\,t)\right) \qquad (2.41)$$

for zero initial conditions.

We attempt now to extend this result to the case in which the forcing frequency ν coincides the natural frequency ω, a case referred to as ***resonance***. Direct calculations are not possible since $x(t)$ for $\nu = \omega$ is indeterminate. We eliminate this indetermination by taking the limit as $\nu \to \omega$ for a fixed but arbitrary time t, i.e.,

$$\lim_{\nu\to\omega}\frac{x_{\,\mathrm{st}}}{1-(\nu/\omega)^2}\left(\sin(\nu\,t) - \frac{\nu}{\omega}\,\sin(\omega\,t)\right) = x_{\,\mathrm{st}}\,\lim_{\nu\to\omega}\frac{t\,\cos(\nu\,t) - \sin(\omega\,t)/\omega}{-2\,\nu/\omega^2}$$
$$= x_{\mathrm{st}}\,\frac{\sin(\omega\,t) - \omega\,t\,\cos(\omega\,t)}{2}$$

by l'Hospital rule (the second limit is obtained by taking the derivatives with respect to ν of the numerator and the denominator of the fraction in the first limit, see Appendix A) so that the displacement at resonance ($\nu = \omega$) has the expression

$$x(t) = \frac{x_{\mathrm{st}}}{2}\big(\sin(\omega\,t) - \omega\,t\,\cos(\omega\,t)\big) = \frac{x_{\mathrm{st}}}{2}\,\sqrt{1+(\omega\,t)^2}\,\sin\left(\omega\,t - \varphi^*\right) \qquad (2.42)$$

by considerations as in Sect. 2.4.2, where $\tan(\varphi^*) = \omega\,t$.

The latter form of $x(t)$ shows that the deformation $x(t)$ increases in time and that the increase is linear in ωt for $\omega t \gg 1$. It also shows that resonance does not mean instant unbounded response. The response is finite in finite time intervals. Yet, it increases in time and will eventually exceed the system strength.

2.4.5.2 Beating Phenomenon

Consider the oscillator and forcing function of the previous section and assume that $\nu \neq \omega$ but the difference between these frequencies is small, i.e., $|\nu - \omega| \ll 1$.

The displacement $x(t)$ of Eq. 2.41 is valid for $\nu \neq \omega$ and can be written in the form

$$\begin{aligned} x(t) & \\ &= \frac{x_{st}}{\omega\left(1-(\nu/\omega)^2\right)}\left[\frac{\omega+\nu}{2}\left(\sin(\nu t)-\sin(\omega t)\right)+\frac{\omega-\nu}{2}\left(\sin(\nu t)+\sin(\omega t)\right)\right] \\ &\simeq \frac{x\ \mathrm{st}}{\omega\left(1-(\nu/\omega)^2\right)}\,\omega\, 2\,\sin\left(\frac{(\nu-\omega)\,t}{2}\right)\cos\left(\frac{(\nu+\omega)\,t}{2}\right), \end{aligned}$$

where the latter expression results by neglecting the second term in the square bracket since $|\nu - \omega| \ll 1$ and by using the trigonometric identity

$$\sin(\nu t) - \sin(\omega t) = 2\,\sin\left((\nu-\omega)\,t)/2\right)\,\cos\left((\nu+\omega)\,t)2\right).$$

This approximation gives

$$x(t) \simeq \left[\frac{2\,x_{st}}{1-(\nu/\omega)^2}\,\sin\left(\frac{(\nu-\omega)\,t}{2}\right)\right]\cos(\omega t) \tag{2.43}$$

and shows that the high frequency component $\cos(\omega t)$ with period $2\pi/\omega$ is modulated by the slow varying amplitude, the term in the square bracket, with period $4\pi/|\nu - \omega|$.

Example 2.11 An undamped oscillator with unit mass and natural frequency $\omega = 6$ is at rest at the initial time and is subjected to an harmonic force $f(t) = q\ \sin(\nu t)$. The left panel of Fig. 2.9 shows the evolution in time of $x(t)$ for $\nu = \omega = 6$ so that system is at resonance. The rate of increase of the amplitude of x is linear in time in agreement with our previous comments. The right panel of Fig. 2.9 shows the displacement $x(t)$ for $\nu = 5.9$ which differs slightly from $\omega = 6$. The plot illustrates the beating phenomenon. Note that the half of the long period of $x(t)$ is slightly larger than 60 in agreement with the period of the square bracket in Eq. 2.43 which is $4\pi/|\nu - \omega| = 125$. The small period is $2\pi/6 \simeq 1$.

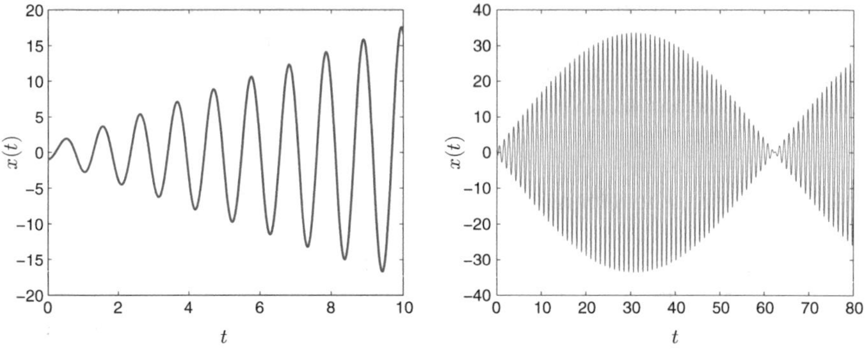

Fig. 2.9 Displacement function $x(t)$ for an undamped SDOF system with natural frequency $\omega = 6$ subjected to a harmonic force with frequency $\nu = \omega = 6$ (left panel) and $\nu = 5.9$ (right panel)

2.4.6 Harmonic Force, Damped System

We have seen that the particular solution for undamped oscillators subjected to harmonic forcing functions can be obtained simply. This section constructs the particular solution $x_p(t)$ for damped SDOF systems subjected to harmonic forces. The Duhamel integral can be used to find $x_p(t)$. The reader is encouraged to implement this approach.

A direct approach is presented here, i.e., we postulate a functional form for $x_p(t)$ and require that it satisfies the differential equation $\ddot{x}_p(t)+2\,\zeta\,\omega\,\dot{x}_p(t)+\omega^2\,x_p(t) = \big(q/m\big)\,\sin(\nu\,t)$. It is shown that the particular solution is

$$x_p(t) = x_{\text{st}}\,r_d(\nu)^2\big[-\big(2\,\zeta\,\nu/\omega\big)\,\cos(\nu\,t) + \big(1-(\nu/\omega)^2\big)\,\sin(\nu\,t)\big], \quad \text{where}$$
$$r_d(\nu) = \frac{1}{\sqrt{\big((1-(\nu/\omega)^2)\big)^2 + \big(2\,\zeta\,\nu/\omega\big)^2}} \quad \text{and } x_{\text{st}} = q/k \tag{2.44}$$

or

$$x_p(t) = x_{\text{st}}\,r_d(\nu)\,\sin\big(\nu\,t-\varphi\big), \quad \text{where } \tan(\varphi) = \frac{2\,\zeta\,\nu/\omega}{1-(\nu/\omega)^2}, \tag{2.45}$$

where $r_d(\nu)$ is called ***dynamic amplification factor*** (DAF).

The particular solution of the previous example does not satisfy the equation of motion since $2\,\zeta\,\omega\,\dot{x}_p(t) = 2\,\zeta\,\omega\,\alpha\,\nu\,\cos\big(\nu\,t\big)$ is the only term which is proportional to $\cos\big(\nu\,t\big)$ so that the equation of motion is satisfied if and only if $2\,\zeta\,\omega\,\alpha\,\nu = 0$ which implies $\alpha = 0$, i.e., $x_p(t) = 0$. Accordingly, the equation of motion is not satisfied since its left and right sides are zero and $\big(q/m\big)\,\sin(\nu\,t)$.

This observation suggests to augment the trial solution of the previous example to

$$x_p(t) = \alpha \sin\left(\nu t\right) + \beta \cos\left(\nu t\right), \quad t \geq 0, \tag{2.46}$$

where α and β are constants which need to be determined. The equation of motion with $x_p(t)$ in place of $x(t)$ has the form

$$\left[\cdots\right] \sin\left(\nu t\right) + \left[\cdots\right] \cos\left[\nu t\right] = (q/m) \sin(\nu t),$$

where the square brackets depend on α and β. For $x_p(t)$ to be a particular solution, the first and second square brackets must be equal to q/m and zero. These conditions give a system of equations for α and β. Following are some computational details.

The equation of motion with $x_p(t)$ in place of $x(t)$ gives

$$\begin{aligned}
&-\alpha \nu^2 \cos(\nu t) - \beta \nu^2 \sin(\nu t) + 2\zeta\omega\left(-\alpha \nu \sin(\nu t) + \beta \nu \cos(\nu t)\right) \\
&+\omega^2 \left(\alpha \cos(\nu t) + \beta \sin(\nu t)\right) = (q/m) \sin(\nu t) \quad \text{or, equivalently,} \\
&\left(-\alpha \nu^2 + 2\zeta\omega\beta\nu + \alpha \nu^2\right) \cos(\nu t) + \left(-\beta \nu^2 - 2\zeta\omega\alpha\nu + \beta \nu^2\right) \\
&\sin(\nu t) = (q/m) \sin(\nu t).
\end{aligned}$$

Since the trial $x_p(t)$ has to satisfy the equation of motion at all times, we require

$$\begin{aligned}
-\alpha \nu^2 + 2\zeta\omega\beta\nu + \alpha \nu^2 &= 0 \quad \text{and} \\
-\beta \nu^2 - 2\zeta\omega\alpha\nu + \beta \nu^2 &= q/m.
\end{aligned}$$

These conditions define a linear system of equations for the unknown coefficients α and β. The solution of this system gives the particular solution of Eq. 2.44.

Similar calculations show that the particular solution for the forcing function $f(t) = q \cos(\nu t)$ has the expression

$$x_p(t) = x_{\text{st}}\, r_d(\nu) \cos(\nu t - \varphi) \tag{2.47}$$

with the notations of Eqs. 2.44 and 2.45 (see Problem 2.5).

We conclude that the particular solutions for the forcing functions $f(t) = \sin(\nu t)$ and $f(t) = \cos(\nu t)$ are

$$\begin{aligned}
f(t) = \sin(\nu t) &\Longrightarrow x_p(t) = \left(r_d(\nu)/\omega^2\right) \sin(\nu t - \varphi) \\
f(t) = \cos(\nu t) &\Longrightarrow x_p(t) = \left(r_d(\nu)/\omega^2\right) \cos(\nu t - \varphi)
\end{aligned} \tag{2.48}$$

since $\sin(\nu t) = (m/q)\left[(q/m) \sin(\nu t)\right]$ and the system is linear so that the particular solution of Eq. 2.45 can be scaled by (q/m), which gives

$$\begin{aligned}
(m/q)\left[x_{\text{st}}\, r_d(\nu) \sin(\nu t - \varphi)\right] &= (m/q)\,(q/k)\, r_d(\nu) \sin(\nu t - \varphi) \\
&= (1/\omega^2)\, r_d(\nu) \sin(\nu t - \varphi).
\end{aligned}$$

2.4.6.1 Alternative Formulation

The calculations simplify significantly if the complex-value representations are used for the force and the particular solution. Set $f(t) = q \exp(i\,\nu\,t)$, where the amplitude q is real-valued as above so that the real and imaginary parts of the forcing function $f(t)$ are $q\,\cos(\nu\,t)$ and $q\,\sin(\nu\,t)$ as $\exp(i\,\nu\,t) = \cos(\nu\,t) + i\,\sin(\nu\,t)$. The complex-valued function $x_p(t) = a(\nu)\exp(i\,\nu\,t)$ is a particular solution provided the complex-valued amplitude $a(\nu)$ satisfies the condition

$$a(\nu)\,e^{i\,\nu\,t}\left((i\,\nu)^2 + 2\,i\,\zeta\,\omega\,\nu + \omega^2\right) = \frac{q}{m}\,e^{i\,\nu\,t},$$

which results by requiring that $x_p(t)$ satisfies the equation of motion at all times. This implies

$$\begin{aligned}a(\nu) &= \frac{q}{m}\,\frac{(\omega^2-\nu^2) - 2\,i\,\omega\,\nu}{(\omega^2-\nu^2)^2 + (2\,\omega\,\nu)^2} = \frac{q}{m\,\omega^2}\,\frac{1-(\nu/\omega)^2 - 2\,i\,\zeta\,\nu/\omega}{\left(1-(\nu/\omega)^2\right)^2 + \left(2\,\zeta\,\nu/\omega\right)^2}\\ &= x_{\,\mathrm{st}}\,r_d(\nu)^2\left(1-(\nu/\omega)^2 - 2\,i\,\zeta\,\nu/\omega\right),\end{aligned} \tag{2.49}$$

where $x_{\mathrm{st}} = q/\left(m\,\omega^2\right)$. The particular solution becomes

$$\begin{aligned}&x_p(t)\\ &= a(\nu)\,e^{i\,\nu\,t} = x_{\mathrm{st}}\,r_d(\nu)^2\left(1-(\nu/\omega)^2 - 2\,i\,\zeta\,\nu/\omega\right)\left(\cos(\nu\,t) + i\,\sin(\nu\,t)\right)\\ &= x_{\,\mathrm{st}}\,r_d(\nu)\left[\left(1-(\nu/\omega)^2\right)r_d(\nu) - \left(2\,i\,\zeta\,\nu/\omega\right)r_d(\nu)\right]\left(\cos(\nu\,t) + i\,\sin(\nu\,t)\right)\\ &= x_{\mathrm{st}}\,r_d(\nu)\left(\cos(\varphi) - i\,\sin(\varphi)\right)\left(\cos(\nu\,t) + i\,\sin(\nu\,t)\right) = x_{\mathrm{st}}\,r_d(\nu)\,e^{-i\,\varphi}\,e^{i\,\nu\,t}\\ &= x_{\mathrm{st}}\,r_d(\nu)\,e^{i\,(\nu\,t-\varphi)} = x_{\mathrm{st}}\,r_d(\nu)\left(\cos(\nu\,t-\varphi) + i\,\sin(\nu\,t-\varphi)\right)\end{aligned} \tag{2.50}$$

where $\sin(\varphi) = 2\,\zeta\,(\nu/\omega)\,r_d(\nu)$ and $\cos(\varphi) = \left[1-(\nu/\omega)^2\right]r_d(\nu)$ so that $\tan(\varphi)$ is as in Eq. 2.45. This shows that the steady-state response to the input $(q/m)\exp(i\,\nu\,t)$ is

$$x_p(t) = x_{\mathrm{st}}\,r_d(\nu)\,\exp\left(i\,(\nu\,t-\varphi)\right)$$

so that the steady-state response to $\exp(i\,\nu\,t)$ is

$$x_p(t) = (m/q)\,x_{\,\mathrm{st}}\,r_d(\nu)\,\exp\left(i\,(\nu\,t-\varphi)\right) = \left(r_d(\nu)/\omega^2\right)\exp\left(i\,(\nu\,t-\varphi)\right)$$

since $(m/q)\,x_{\mathrm{st}} = (m/q)\,(q/k) = m/k = 1/\omega^2$.

To summarize, the steady-state response to the input $\exp(i\,\nu\,t)$ in Eq. 2.50 has the same frequency as the input but different size and phase. The input-output relationship is

$$f(t) = e^{i\,\nu\,t} \Longrightarrow x_p(t) = \left(r_d(\nu)/\omega^2\right) e^{i\,(\nu\,t-\varphi)} \quad \text{which implies}$$
$$f(t) = \cos(\nu\,t) \Longrightarrow x_p(t) = \left(r_d(\nu)/\omega^2\right) \cos(\nu\,t-\varphi) \quad \text{and}$$
$$f(t) = \sin(\nu\,t) \Longrightarrow x_p(t) = \left(r_d(\nu)/\omega^2\right) \sin(\nu\,t-\varphi) \tag{2.51}$$

by the linearity of the equation of motion. These particular solutions are the steady-state solutions to the inputs $f(t) = \exp(i\,\nu\,t)$, $f(t) = \cos(\nu\,t)$, and $f(t) = \sin(\nu\,t)$ since the general solution of the homogeneous equation vanishes for large times (see Eq. 2.13).

2.4.6.2 Dynamic Amplification Factor (DAF)

The form of the particular solution $x_p(t) = x_{\text{st}}\,r_d(\nu)\,\sin(\nu\,t-\varphi)$ in Eq. 2.45 shows that the dynamic amplification factor $r_d(\nu)$ gives the scale of the dynamic response relative to the static response x_{st}, see Fig. 2.10 which shows $r_d(\nu)$ for two values of the damping ratio ζ over a broad range of ν/ω ratios.

The ratio ν/ω controls not only the scale of the particular solution but also its phase. We consider three cases, $\nu/\omega \ll 1$, $\nu/\omega \gg 1$ and $\nu/\omega \simeq 1$. If $\nu/\omega \ll 1$, the DAF $r_d(\nu) \simeq 1$ and $\tan(\varphi) \simeq 0$ which implies $\varphi \simeq 0$ so that $x_p(t)$ describes a quasi-static response in phase with the forcing function. If $\nu/\omega \gg 1$, the DAF $r_d(\nu) \sim O(\omega/\nu)^2$ and $\tan(\varphi) \to 0$ through negative values as $\nu/\omega \to \infty$. This means that, for large values of ν/ω, the DAF is small, and so is the system response $x_p(t)$, and $\varphi \simeq \pi$, i.e., the solution $x_p(t)$ and the forcing functions are out of phase.

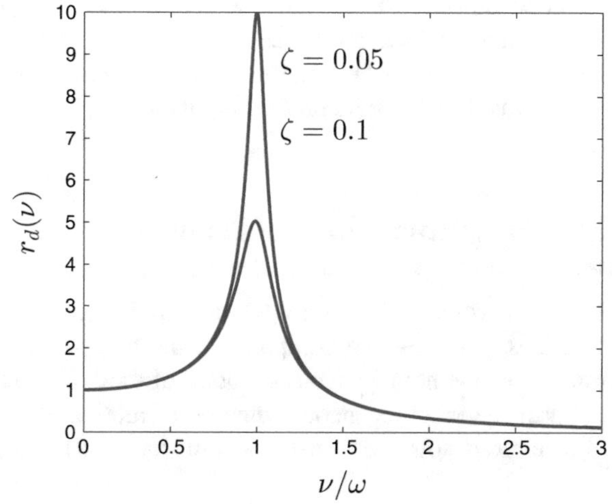

Fig. 2.10 DAF $r_d(\nu)$ for $\zeta = 0.05$ and 0.1

Fig. 2.11 Particular and general solutions $x_p(t)$ and $x(t)$ (solid and dashed lines) for $\omega = 6$, $\zeta = 0.1$, $\nu = 5$, $x_0 = 2$, $\dot{x}_0 = 6$, and $x_{st} = 0.3$

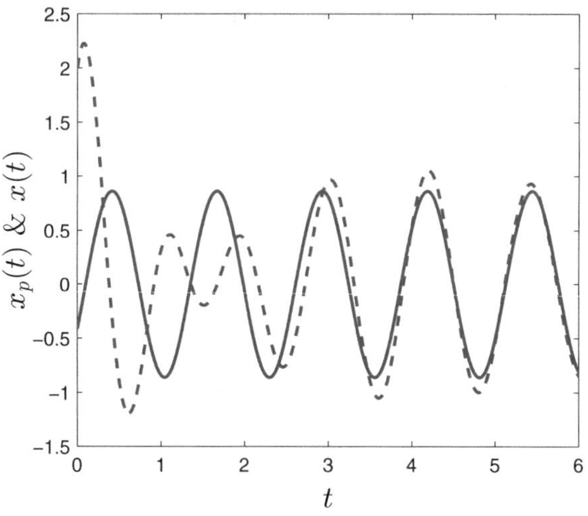

If $\nu/\omega \simeq 1$, the DAF takes nearly its largest value and the phase φ of $x_p(t)$ relative to the forcing function is $\pm\pi/2$ since $\tan(\varphi) \to \pm\infty$ as $\nu/\omega \to 1$ from the left ($\nu < \omega$) and from the right ($\nu > \omega$).

Example 2.12 Consider a SDOF system with natural frequency $\omega = 6$ and damping ratio $\zeta = 0.1$ which is subjected to a harmonic forcing function $f(t) = q \sin(\nu t)$ with $\nu = 5$ and q such that $x_{st} = q/k = 0.3$. The solid line in Fig. 2.11 is the particular solution $x_p(t)$ given by Eq. 2.44. The dashed line is the general solution

$$x(t) = e^{-\zeta \omega t} \left(A \cos(\omega_d t) + B \sin(\omega_d t)\right) + x_p(t)$$

whose constants are selected to satisfy the initial conditions $x_0 = 2$ and $\dot{x}_0 = 6$.

The visual inspection of the figure shows that the general solutions satisfy the initial condition $x(0) = 2$ and that $\dot{x}(0) > 0$. Its value matches the initial condition $\dot{x}_0 = 6$ although it cannot be seen at the scale of the figure. Note also that the general solution $x(t)$ approaches the steady-state solution $x_{ss}(t) = x_p(t)$ as time increases in agreement with our theoretical arguments.

2.5 Frequency Domain Analysis

The displacement function $x(t)$ of SDOF systems subject to arbitrary forcing functions $f(t)$ has two components, the free vibration solution, i.e., the general solution of the homogeneous equation of motion $x_h(t)$, and the particular solution $x_p(t)$ of the inhomogeneous equation of motion. For damped SDOF systems, the free vibration solution vanishes as time increases so that $x(t)$ can be approximated by the steady-state solution, i.e., $x(t) \simeq x_{ss}(t) = x_p(t)$, for sufficiently large times. The focus of the frequency domain analysis is on the steady-state solution.

We first represent the forcing functions $f(t)$ by sums of harmonics of different frequencies and amplitudes and then construct similar representations for the oscillator steady-state displacement. We show that the input $f(t)$ and the output $x_{ss}(t)$ have the same frequencies but different amplitudes. The output amplitudes are controlled by DAFs.

2.5.1 Fourier Series Representation of Forcing Functions

This section provides a brief review of essentials on Fourier series which are needed to find the response of SDOF systems in the frequency domain. Detailed technical considerations can be found in [4] (Chaps. 1–3).

Suppose that the applied force $f(t)$ is a periodic function with period τ, i.e., $f(t) = f(t+\tau)$ for any t in the domain of definition of $f(t)$, and consider the representation

$$f(t) = \frac{a_0}{2} + \sum_{i=1}^{\infty} \big(a_i \cos(\nu_i t) + b_i \sin(\nu_i t)\big), \tag{2.52}$$

where $\nu_1 = 2\pi/\tau$ and $\nu_i = i\,\nu_1$, $i = 1, 2, \ldots$, are multiple of the fundamental frequency ν_1.

To determine the coefficients (a_i, b_i) in Eq. 2.52, we assume that the series representation of $f(t)$ can be integrated term by term. Under this assumption, the integrals over $[0, \tau]$ of the both sides of Eq. 2.52 multiplied by 1, $\cos(\nu_j t)$ and $\sin(\nu_j t)$, $j = 1, 2, \ldots$, give

$$\begin{aligned} a_0 &= \frac{2}{\tau}\int_0^{\tau} f(t)\,dt, \\ a_j &= \frac{2}{\tau}\int_0^{\tau} f(t)\cos(\nu_j t)\,dt, \quad j \geq 1 \\ b_j &= \frac{2}{\tau}\int_0^{\tau} f(t)\sin(\nu_j t)\,dt \quad j = 1, 2, \ldots, \end{aligned} \tag{2.53}$$

by using the orthogonality of the trigonometric functions, i.e.,

$$\begin{aligned} &\int_0^{\tau} \cos(\nu_i t)\cos(\nu_j t)\,dt = \frac{\tau}{2}\,\delta_{ij}, \\ &\int_0^{\tau} \sin(\nu_i t)\sin(\nu_j t)\,dt = \frac{\tau}{2}\,\delta_{ij}, \\ &\int_0^{\tau} \sin(\nu_i t)\cos(\nu_j t)\,dt = 0, \end{aligned} \tag{2.54}$$

where $\delta_{ij} = 1$ for $i = j$ and zero otherwise. The coefficients of Eq. 2.53 are called the ***Fourier coefficients*** of $f(t)$ and the representation of Eq. 2.52 with these coefficients is referred to as the ***Fourier series*** of $f(t)$.

Consider the truncated version

$$f_n(t) = \frac{a_0}{2} + \sum_{i=1}^{n} \big(a_i \cos(\nu_i t) + b_i \sin(\nu_i t)\big), \quad n = 1, 2, \ldots, \tag{2.55}$$

of the Fourier series of $f(t)$ which is used in numerical calculations. The accuracy of this representation of $f(t)$ depends on the truncation level. The following paragraph provides some technical comments on the accuracy of $f_n(t)$ and the validity of the calculations in Eq. 2.53 involving term by term integration. The reader may skip these comments unless interested in technicalities.

It can be shown that f_n converges uniformly to f on $[0, \tau]$, i.e., $\max_{0\le t\le\tau} |f_n(t) - f(t)|$ becomes small and remains small from an index n on, i.e., given $\varepsilon > 0$, there is n_ε such that $\max_{0\le t\le\tau} |f_n(t) - f(t)| \le \varepsilon$ for $n \ge n_\varepsilon$. If $f(t)$ has discontinuities, $f_n(t)$ converges to the arithmetic mean of the right and left limits $\big(f(t+) - f(t-)\big)/2$ of $f(t)$ at its discontinues points [4] (Sect. 1.10). It can also be shown that the term by term integration which allows us to calculate the Fourier coefficient of the Fourier series of $f(t)$ is valid for continuous, piecewise differentiable functions [4] (Sect. 3.10).

Example 2.13 The solid lines in Fig. 2.12 are the graph of a target function $f(t)$ with the support $[0, \tau = 10]$. The function is continuous but is not periodic in this time interval as $f(0) \neq f(\tau)$. The top left, top right, bottom left, and bottom right panels show with dashed lines the truncated Fourier series representations $f_n(t)$ for $n = 0$, $n = 5$, $n = 10$, and $n = 30$. Visual inspection suggests that the accuracy of $f_n(t)$ improves with n. As expected, the representations $f_n(t)$ are in error at the ends of the interval $[0, \tau]$ since $f(t)$ is not periodic.

Considerations as in Sect. 2.4.2.2 can be used to construct the following amplitude-phase representation of $f_n(t)$

$$f_n(t) = \frac{a_0}{2} + \sum_{i=1}^{n} \sqrt{a_i^2 + b_i^2}\, \sin(\nu_i t + \varphi_i), \quad n = 1, 2, \ldots, \tag{2.56}$$

where $\tan(\varphi_i) = a_i/b_i$. This follows from the equalities

$$\sqrt{a_i^2 + b_i^2}\left(\frac{a_i}{\sqrt{a_i^2 + b_i^2}} \cos(\nu_i t) + \frac{b_i}{\sqrt{a_i^2 + b_i^2}} \sin(\nu_i t)\right)$$

$$= \sqrt{a_i^2 + b_i^2}\Big(\sin(\varphi_i) \cos(\nu_i t) + \cos(\varphi_i) \sin(\nu_i t)\Big) = \frac{b_i}{\sqrt{a_i^2 + b_i^2}} \sin(\nu_i t + \varphi_i),$$

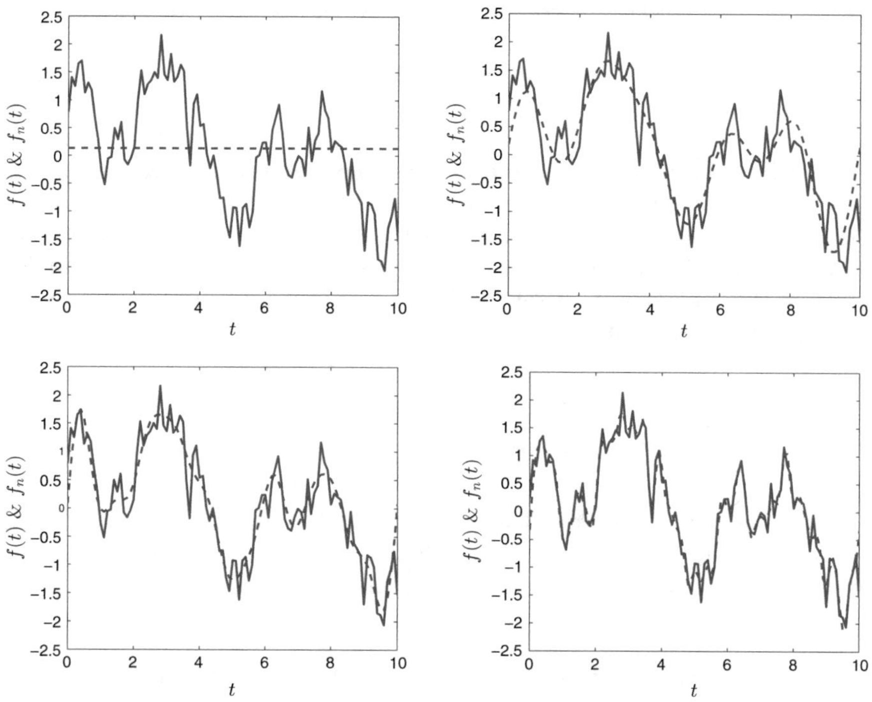

Fig. 2.12 Solid and dashed lines are target function $f(t)$ and truncated Fourier series $f_n(t)$ representations for $n = 0$, $n = 5$, $n = 10$, and $n = 30$ (top left, top right, bottom left, and bottom right panels)

where $\sin(\varphi_i) = a_i/\sqrt{a_i^2 + b_i^2}$ and $\cos(\varphi_i) = b_i/\sqrt{a_i^2 + b_i^2}$. We consider a simplified version of the Fourier transform which retains information on only the signal frequencies and amplitudes. These Fourier transforms of the Fourier series representations of Eqs. 2.52 and 2.56 have the form

$$\text{FT}[f](\nu) = \frac{a_0}{2}\,\delta(\nu) + \sum_{i=1}^{\infty} \sqrt{a_i^2 + b_i^2}\,\delta(\nu - \nu_i) \quad \text{and}$$

$$\text{FT}[f_n](\nu) = \frac{a_0}{2}\,\delta(\nu) + \sum_{i=1}^{n} \sqrt{a_i^2 + b_i^2}\,\delta(\nu - \nu_i), \tag{2.57}$$

where $\delta(\nu - \nu_i) = 1$ for $\nu = \nu_i$ and zero otherwise.

We conclude our brief considerations on Fourier series with the observation that the Fourier series of $f(t)$ can be viewed as an element of the linear space spanned by the orthogonal system of functions

$$\{1, \cos(\nu_1\, t), \sin(\nu_1\, t), \cdots, \cos(\nu_i\, t), \sin(\nu_i\, t), \cdots\}$$

and that the Fourier coefficients are projections of $f(t)$ on these basis functions (see Appendix C). The Fourier series of $f(t)$ is conceptually similar to the representation of vectors $\mathbf{v}$ in, e.g., the three-dimensional Euclidian space, which has the form $\mathbf{v} = v_x\,\mathbf{i} + v_y\,\mathbf{j} + v_z\,\mathbf{k}$, where $\mathbf{i}, \mathbf{j}, \mathbf{k}$ denote the unit vectors along the x, y, z-coordinates and v_x, v_y, v_z are the projections of $\mathbf{v}$ on these unit vectors. Note that the representation of $\mathbf{v}$ and $f(t)$ have the same form. They are sums of projections of these elements on the orthogonal systems spanning their spaces. Note also that approximations of $f(t)$ by its values at $N < \infty$ times in $[0, \tau]$ are vectors in N-dimensional Euclidian spaces.

2.5.2 Steady-State Solution

Consider the oscillator in Eq. 2.8 and approximate the forcing function $f(t)$ by its truncated Fourier series $f_n(t)$ of Eq. 2.56 so that the oscillator displacement $x(t)$ is approximated by the solution $x_n(t)$ of the differential equation

$$\ddot{x}_n(t) + 2\,\zeta\,\omega\,\dot{x}_n(t) + \omega^2\,x_n(t) = \frac{1}{m}\left[\frac{a_0}{2} + \sum_{i=1}^{n}\Big(a_i\,\cos(\nu_i\,t) + b_i\,\sin(\nu_i\,t)\Big)\right]. \tag{2.58}$$

The initial conditions are not specified since we are interested in the steady-state solution of this equation.

Since the system is linear, $x_n(t)$ can be constructed by adding the solutions of the above equation to the forcing functions $a_0/(2\,m)$, $\{(a_i/m)\,\cos(\nu_i\,t)\}$, and $\{(b_i/m)\,\sin(\nu_i\,t)\}$. This statement also follows from the Duhamel integral of Eq. 2.30 which gives

$$x_n(t) = \int_0^t h(t-u)\,f_n(u)\,du = \frac{1}{m}\left[\frac{a_0}{2}\int_0^t h(t-u)\,du \right.$$
$$\left. + \sum_{i=1}^{n}\left(a_i\int_0^t h(t-u)\,\cos(\nu_i\,u)\,du + b_i\int_0^t h(t-u)\,\sin(\nu_i\,u)\,du\right)\right]$$

since the integral is a linear operator.

We have already calculated the steady-state solutions to the above forcing functions, see Eq. 2.14, Eq. 2.48 and Eq. 2.51. These input-output relationships applied to the terms in the representation of $f_n(t)/m$ are

$$a_0/(2\,m) \Longrightarrow a_0/(2\,m\,\omega^2)$$
$$(a_i/m)\,\cos(\nu_i\,t) \Longrightarrow (a_i/m)\,\big(r_d(\nu_i)/\omega^2\big)\,\cos(\nu_i\,t - \varphi_i)$$

$$(b_i/m)\,\sin(\nu_i\,t) \Longrightarrow (b_i/m)\left(r_d(\nu_i)/\omega^2\right)\,\sin(\nu_i\,t-\varphi_i)$$

so that the steady-state solution $x_n(t)$ is

$$x_n(t) = \frac{a_0}{2\,m\,\omega^2} + \sum_{i=1}^{n}\left[\frac{a_i}{m}\,\frac{r_d(\nu_i)}{\omega^2}\,\cos(\nu_i\,t-\varphi_i) + \frac{b_i}{m}\,\frac{r_d(\nu_i)}{\omega^2}\,\sin(\nu_i\,t-\varphi_i)\right] \tag{2.59}$$

which has the amplitude-phase representation

$$\begin{aligned} x_n(t) &= \frac{a_0}{2\,m\,\omega^2} + \sum_{i=1}^{n}\frac{r_d(\nu_i)}{m\,\omega^2}\,\sqrt{a_i^2+b_i^2} \\ &\quad\times\left[\frac{a_i}{\sqrt{a_i^2+b_i^2}}\,\cos(\nu_i\,t-\varphi_i) + \frac{b_i}{\sqrt{a_i^2+b_i^2}}\,\sin(\nu_i\,t-\varphi_i)\right] \\ &= \frac{a_0}{2\,m\,\omega^2} + \sum_{i=1}^{n}\frac{r_d(\nu_i)}{m\,\omega^2}\,\sqrt{a_i^2+b_i^2}\,\sin(\nu_i\,t-\varphi_i+\theta_i) \end{aligned} \tag{2.60}$$

by using the notations

$$\sin(\theta_i) = \frac{a_i}{\sqrt{a_i^2+b_i^2}} \quad \text{and} \quad \cos(\theta_i) = \frac{b_i}{\sqrt{a_i^2+b_i^2}} \tag{2.61}$$

and the equality $\sin(\theta_i)\,\cos(\nu_i\,t-\varphi_i)+\cos(\theta_i)\,\sin(\nu_i\,t-\varphi_i) = \sin(\nu_i\,t-\varphi_i+\theta_i)$. The Fourier transform of the steady-state solution $x_n(t)$ has the expression

$$\mathrm{FT}[x_n](\nu) = \frac{a_0}{2\,m\,\omega^2}\,\delta(\nu) + \sum_{i=1}^{n}\frac{r_d(\nu_i)}{m\,\omega^2}\,\sqrt{a_i^2+b_i^2}\,\delta(\nu-\nu_i). \tag{2.62}$$

The comparison of the Fourier transforms of the forcing function $f_n(t)$ and the steady-state solution $x_n(t)$ of Eqs. 2.57 and 2.62 shows that:

1. The forcing function $f_n(t)$ and the steady-state solution $x_n(t)$ have the same frequency content, i.e., the functions $f_n(t)$ and $x_n(t)$ are sums of harmonics of frequencies $\nu_0 = 0$ and ν_i, $i = 1, \ldots, n$.
2. The difference between the Fourier transforms of $f_n(t)$ and $x_n(t)$ is the amplitudes and the phases of their constitutive harmonics, $a_0/2$; $\sqrt{a_i^2+b_i^2}$ and $\{\varphi_i\}$ for $f_n(t)$ and $a_0/(2\,m\,\omega^2) = a_0/(2\,k)$; $r_d(\nu_i)/(m\,\omega^2)\sqrt{a_i^2+b_i^2} = (r_d(\nu_i)/k)\sqrt{a_i^2+b_i^2}$ and $\{\theta_i-\varphi_i\}$ for $x_n(t)$.
3. The relationship

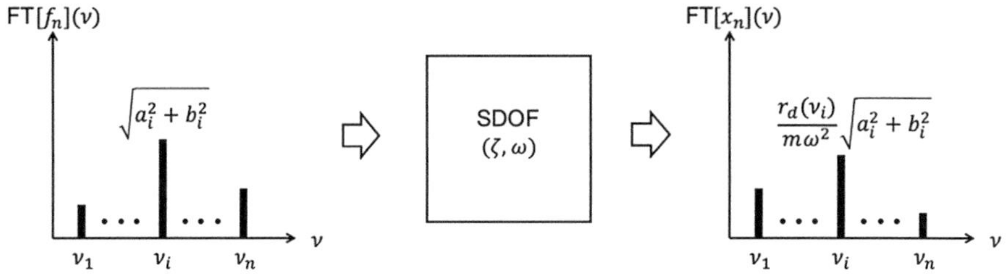

Fig. 2.13 Fourier transforms of forcing function and steady-state displacement

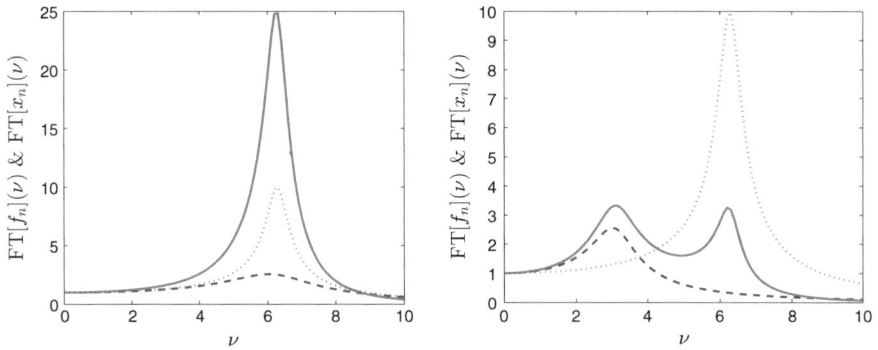

Fig. 2.14 Input and output Fourier transforms (dashed and solid lines) for ω under the peak of FT[f_n](ν) and away from it (left and right panels). The dotted lines are the DAF for two designs, i.e., two natural frequencies

$$\mathrm{FT}[x_n](\nu_i) = \frac{r_d(\nu_i)}{m\,\omega^2}\,\mathrm{FT}[f_n](\nu_i) = \frac{r_d(\nu_i)}{k}\,\mathrm{FT}[f_n](\nu_i), \quad i = 1, \ldots, n, \tag{2.63}$$

between the Fourier transforms of $f_n(t)$ and $x_n(t)$ is illustrated in Fig. 2.13 for ν_i, $i = 1, \ldots, n$. The output amplitudes are scaled versions of input amplitudes via the dynamic amplification factor $r_d(\nu)$. The constant $a_0/2$ in the Fourier transform of $f_n(t)$ is mapped into $a_0/(2\,k)$ in the Fourier transform of the response since there is no dynamic amplification for $\nu = 0$ ($r_d(0) = 1$).

4. The relationship of Eq. 2.63 can be used directly to optimize the design of SDOF systems, as illustrated by the cartoon in Fig. 2.14 which shows the Fourier transforms of $x_n(t)$ and $f_n(t)$ (with heavy solid and dashed lines) and the dynamic amplification factor (dotted lines). The system response in the left panel is large since the input Fourier transform FT[f_n](ν) and the dynamic amplification factor $r_d(\nu)$ peak at the same frequency. The system response in the right panel is much smaller since the maxima of the input Fourier transform FT[f_n](ν) and the dynamic amplification factor $r_d(\nu)$ occur at different frequencies. The change from the design in the left panel to that in the right panel can be accomplished by changing, e.g., the oscillator stiffness.

Example 2.14 A damped single degree of freedom system with natural frequency $\omega = 3\pi/2$ rad/sec and damping ratio $\zeta = 0.02$ is subjected to a seismic ground acceleration

$$a(t) = \sum_{i=1}^{n} [a_i \cos(\nu_i t) + b_i \sin(\nu_i t)], \quad 0 \le t \le \tau,$$

where $n = 10$, $\nu_i = 2\pi i/\tau$, $\tau = 10$ sec, $a_i = \sin(\pi/i)$, and $b_i = \sin(3\pi/(2i))$. The Fourier transforms of the ground acceleration $a(t)$ and steady-state displacement are

$$\mathrm{FT}[a](\nu) = \sum_{i=1}^{n} \sqrt{a_i^2 + b_i^2}\, \delta(\nu - \nu_i)$$

and

$$\mathrm{FT}[x](\nu) = \sum_{i=1}^{n} \frac{r_d(\nu)}{\omega^2} \sqrt{a_i^2 + b_i^2}\, \delta(\nu - \nu_i).$$

The scaling from $\mathrm{FT}[a(t)](\nu)$ to $\mathrm{FT}[x(t)](\nu)$ is $r_d(\nu)/\omega^2$ rather than $r_d(\nu)/(m\,\omega^2)$ since the forcing function is $-a(t)$ rather than $f(t)/m$ (see Eq. 2.33). The reader is encouraged to construct the input/output Fourier transforms, i.e., the Fourier transforms $\mathrm{FT}[a](\nu)$ and $\mathrm{FT}[x](\nu)$, based on our discussion on Fig. 2.13.

2.6 Numerical Methods

The general solution of the equation of motion given by Eq. 2.8 can be obtained by (1) the numerical integration of the Duhamel integral of Eq. 2.31, (2) the finite difference (FD) or other numerical methods for solving the equation of motion, or (3) the Fourier series representation of the forcing function. Solutions by the FD method and by the Fourier series representation are discussed.

2.6.1 Finite Difference (FD) Method

We construct FD approximations of the derivatives $\dot{x}(t)$ and $\ddot{x}(t)$ of the displacement functions $x(t)$, introduce these approximations in the equation of motion, which becomes a linear system of algebraic equations whose unknowns are displacement at discrete times. The solution of this system of algebraic equations provides an approximation for $x(t)$.

2.6.1.1 FD Approximations of Derivatives

Consider a real-valued function g which is differentiable as many times as needed and use Taylor's formula (see Appendix A) to construct finite difference approximations of this function. For $h > 0$ small and t arbitrary, we have

$$g(t+h) = g(t) + \frac{h}{1!}\, g'(t) + \frac{h^2}{2!}\, g''(t) + \frac{h^3}{3!}\, g'''(t) + O(h^4)$$
$$g(t-h) = g(t) - \frac{h}{1!}\, g'(t) + \frac{h^2}{2!}\, g''(t) - \frac{h^3}{3!}\, g'''(t) + O(h^4), \tag{2.64}$$

where $O(h^4)$ includes term of order h^4 and higher. The addition and the subtraction of the above equations give $g(t+h) + g(t-h) = 2\, g(t) + h^2\, g''(t) + O(h^4)$ and $g(t+h) - g(t-h) = 2\, h\, g'(t) + O(h^3)$ so that

$$g''(t) = \frac{g(t+h) - 2\, g(t) + g(t-h)}{h^2} + O(h^2) \quad \text{and}$$
$$g'(t) = \frac{g(t+h) - g(t-h)}{2\, h} + O(h^2). \tag{2.65}$$

The approximations of the first and second-order derivatives of $g(t)$ in Eq. 2.65, referred to as ***finite difference approximations***, are obtained from the values of the function at t and of left and right of t, i.e., the values $g(t)$, $g(t-h)$ and $g(t+h)$. Moreover, the FD approximations of $g'(t)$ and $g''(t)$ have the same accuracy in the sense that their errors are of order h^2. Similar formulas can be developed for higher order derivatives.

Note also that simpler FD approximations are less accurate, e.g., the forward FD approximation. This approximation can be obtained from, e.g., the first equality of Eq. 2.64 which gives $g(t+h) = g(t) + \frac{h}{1!}\, g'(t) + O(h^2)$, so that the error of the resulting FD approximation

$$g'(t) = \frac{g(t+h) - g(t)}{h} + O(h) \tag{2.66}$$

is of order h rather than h^2 as in Eq. 2.65

2.6.1.2 FD Version of the Equation of Motion

Suppose our objective is to find the FD solution of Eq. 2.8 over a finite time interval for the ICs $(x_0, \dot{x}_0)$. The solution involves the following three steps.

- *Step 1:* Select a relatively small time step $\Delta t > 0$ and denote by $x_k = x(k\,\Delta t)$ and $f_k = f(k\,\Delta t)$, $k = 0, 1, \ldots$, the oscillator displacement and the forcing function at the discrete times $t_k = k\,\Delta t$.

- *Step 2:* Write the FD difference version of the equation of motion at each discrete time t_k by replacing the derivatives in this equation with their FD approximations, i.e.,

$$\frac{x_{k+1} - 2\,x_k + x_{k-1}}{(\Delta t)^2} + 2\,\zeta\,\omega\,\frac{x_{k+1} - x_{k-1}}{2\,\Delta t} + \omega^2\,x_k = \frac{f_k}{m} + O((\Delta t)^2),$$
$$k = 0, 1, 2, \ldots, \tag{2.67}$$

or

$$\left(\frac{1}{(\Delta t)^2} + \frac{\zeta\,\omega}{\Delta t}\right) x_{k+1} + \left(-\frac{2}{(\Delta t)^2} + \omega^2\right) x_k + \left(\frac{1}{(\Delta t)^2} - \frac{\zeta\,\omega}{\Delta t}\right) x_{k-1}$$
$$= \frac{f_k}{m} + O((\Delta t)^2), \tag{2.68}$$

so that we have the recurrence formula

$$x_{k+1} = \alpha\,x_k + \beta\,x_{k-1} + \gamma\,f_k, \quad k = 0, 1, \ldots, \tag{2.69}$$

by neglecting terms of order $O(\Delta t^2)$, where

$$\alpha = -\left(-\frac{2}{(\Delta t)^2} + \omega^2\right)\xi, \quad \beta = -\left(\frac{1}{(\Delta t)^2} - \frac{\zeta\,\omega}{\Delta t}\right)\xi, \quad \gamma = \xi/m,$$
$$1/\xi = \left(\frac{1}{(\Delta t)^2} + \frac{\zeta\,\omega}{\Delta t}\right) \tag{2.70}$$

- *Step 3:* Initiate the recurrence formula of Eq. 2.69 by using the ICs $(x_0, \dot{x}_0)$. For $k = 0$, this formula, $x_1 = \alpha\,x_0 + \beta\,x_{-1} + \gamma\,f_0$, depends on x_0, which is the known initial displacement, and x_{-1}, which is a value of the displacement function $x(t)$ outside its domain of definition. We can find x_{-1} from the FD representation of the initial velocity which gives

$$\dot{x}_0 = (x_1 - x_{-1})/(2\,\Delta t) + O((\Delta t)^2) \quad \text{or } x_{-1} \simeq x_1 - 2\,\Delta t\,\dot{x}_0$$

so that the recurrence formula for $k = 0$ becomes $x_1 \simeq \alpha\,x_0 + \beta\,(x_1 - 2\,\Delta t\,\dot{x}_0) + \gamma\,f_0$ which gives

$$x_1 \simeq \frac{1}{1-\beta}\left(\alpha\,x_0 - 2\,\beta\,\Delta t\,\dot{x}_0 + \gamma\,f_0\right). \tag{2.71}$$

The initial condition x_0 and the displacement x_1 in Eq. 2.71 can be used to initiate the recurrence formula of Eq. 2.69 and calculate the system displacement at the times t_k, $k = 2, 3, \ldots$, over the time interval of interest. We note that commercial codes do not modify the recurrence formula as done above to find

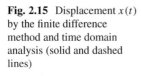

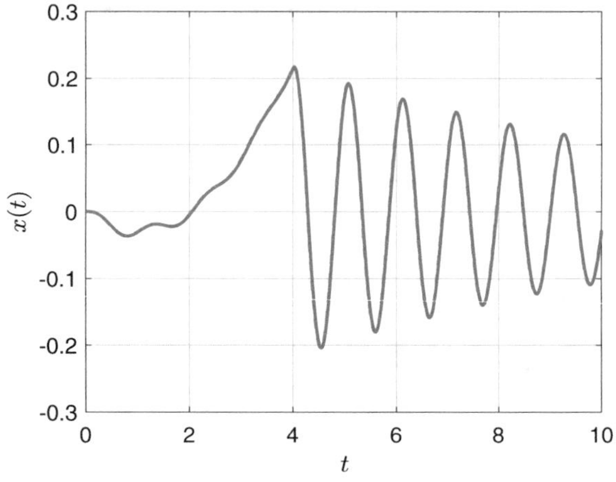

Fig. 2.15 Displacement $x(t)$ by the finite difference method and time domain analysis (solid and dashed lines)

x_1. Instead, they augment the FD difference equations with initial conditions to obtained FD solutions.

Example 2.15 Consider a damped SDOF system with mass $m = 1$ kg, natural frequency $\omega = 6$ rad/s, and damping ratio $\zeta = 2\%$. The oscillator is at rest at the initial time and is subjected to the forcing function $p(t) = t^2 - 2t$ N during the time interval $[0, 4]$ sec. The solid and dashed lines of Fig. 2.15 show the oscillator displacement $x(t)$ during the time interval $[0, 10]$ sec obtained by the finite difference method of this section and the time domain analysis of Sect. 2.4. The two solutions are nearly indistinguishable at the scale of the figure.

The implementation of the FD method has followed the steps outlined in this section. First, a time step of $\Delta t = 0.1$ sec was used to discretize the time interval $[0, 10]$. Second, the recurrence formula of Eq. 2.69 was constructed for this time step. Third, the displacement x_1 at time $t = \Delta t$ was calculated from Eq. 2.71. This displacement and the initial condition $x_0 = 0$ have been used to initiate the recurrence formula and calculate the displacement $x(t)$ at the discrete times of the time range under consideration.

2.6.2 Fourier Series (FS) Method

The displacement of a SDOF subjected to an arbitrary force $f(t)$ can be calculated from Eq. 2.31 and has the form

$$x(t) = e^{-\zeta \omega t} \big(A \cos(\omega_d t) + B \sin(\omega_d t)\big) + \int_0^t h(t-u)\, f(u)\, du, \qquad (2.72)$$

where the constants A and B result from the initial conditions $x(0) = x_0$ and $\dot{x}(0) = \dot{x}_0$.

Suppose that the oscillator is subjected to a large family of forcing functions $\{f_k(t)\}$, $k = 1, \ldots, m$, which are defined on the same time interval $[0, \tau]$. The displacement functions $\{x_k(t)\}$, $k = 1, \ldots, m$, to these family of forces can be obtained by standard numerical algorithms by calling these algorithms m times. A more efficient approach is to calculate the displacements $\{x_k(t)\}$ to Fourier series representations of the forcing functions $\{f_k(t)\}$. The implementation of this method involves the following three steps.

- *Step 1:* Represent the forcing functions $\{f_k(t)\}$ by Fourier series. Since the forces $\{f_k(t)\}$ are active on the same time interval $[0, \tau]$, they admit the representations (see Eq. 2.55)

$$f_{k,n}(t) = \frac{a_{k,0}}{2} + \sum_{i=1}^{n} \big(a_{k,i} \cos(\nu_i t) + b_{k,i} \sin(\nu_i t)\big), \quad n = 1, 2, \ldots, \tag{2.73}$$

where the coefficients $\{a_{k,i}\}$ and $\{b_{k,i}\}$ are given by the formulas of $\{a_i\}$ and $\{b_i\}$ in Eq. 2.53 with f_k in place of f. The truncation level n has to be such that all members of the family of forces $\{f_k(t)\}$ are represented accurately.
- *Step 2:* Calculate and store the integrals

$$I_{c,i}(t) = \int_0^t h(t-u) \cos(\nu_i u)\, du \quad \text{and} I_{s,i}(t) = \int_0^t h(t-u) \sin(\nu_i u)\, du \tag{2.74}$$

for t in $[0, \tau]$. Note that the family of integrals

$$I_{k,n}(t) = \int_0^t h(t-u)\, f_{k,n}(u)\, du = \sum_{i=1}^{n} \big[a_{k,i}\, I_{c,i}(t) + b_{k,i}\, I_{s,i}(t)\big], \quad 0 \le t \le \tau, \tag{2.75}$$

corresponding to the integral of Eq. 2.72 with $\{f_{k,n}\}$ in place of f result by elementary calculations by using the integrals of Eq. 2.74 which have been stored.
- *Step 3:* The displacements $\{x_k(t)\}$ have the form

$$x_k(t) = e^{-\zeta \omega t} \big(A_k \cos(\omega_d t) + B_k \sin(\omega_d t)\big) + I_{k,n}(t), \quad 0 < t < \tau, \tag{2.76}$$

where the constants A_k and B_k can be obtained from the initial conditions x_0 and $\dot{x}_0$. The integrals $\{I_{c,i}(t)\}$ and $\{I_{s,i}(t)\}$ in the expression of $x_k(t)$ are zero at the initial time $t = 0$ and so are their time derivatives

$$I'_{c,i}(t) = \int_0^t \frac{\partial h(t-u)}{\partial t} \cos(\nu_i\, u)\, du + h(0)\, \cos(\nu_i\, t) \Longrightarrow I'_{c,i}(0) = 0 \quad \text{and}$$

$$I'_{s,i}(t) = \int_0^t \frac{\partial h(t-u)}{\partial t} \sin(\nu_i\, u)\, du + h(0)\, \sin(\nu_i\, t) \Longrightarrow I'_{s,i}(0) = 0$$

since $h(0) = 0$. Then, the initial conditions $x_k(0) = x_0$ and $\dot{x}_k(0) = \dot{x}_0$ imply $A_k = x_0$ and $-\zeta\, \omega\, A_k + \omega_d\, B_k = \dot{x}_0$, so that

$$x_k(t) = e^{-\zeta\, \omega t} \left(x_0 \cos(\omega_d\, t) + \frac{\dot{x}_0 + \zeta\, \omega\, x_0}{\omega_d} \sin(\omega_d\, t) \right) + \sum_{i=1}^{n} \left[a_{k,i}\, I_{c,i}(t) + b_{k,i}\, I_{s,i}(t) \right]. \tag{2.77}$$

Since $\{a_{k,i}\}$, $\{b_{k,i}\}$, $\{I_{c,i}(t)\}$, and $\{I_{s,i}(t)\}$ are known, the time histories of the displacements $\{x_k(t)\}$, $k = 1, \ldots, m$, result by elementary calculations.

Example 2.16 Consider an oscillator with natural frequency $\omega = \pi$, damping ratio $\zeta = 0.05$, and mass $m = 1$ which is at rest at the initial time. The system is subjected to a family of two forcing functions $\{f_{k,n}(t)\}$, $k = 1, 2$, given by Eq. 2.73 with $n = 3$, $a_{1,0} = a_{2,0} = 0$, $(a_{1,1}, a_{1,2}, a_{1,3}) = (2, 4, 3)$, $(b_{1,1}, b_{1,2}, b_{1,3}) = (1, 2, 3)$, $(a_{2,1}, a_{2,2}, a_{2,3}) = (-5, -2, 5)$ and $(b_{2,1}, b_{2,2}, b_{2,3}) = (2, 1, 3)$.

The solid and dashed lines in the left panel of Fig. 2.16 are the functions $I_{c,i}(t)$ and $I_{s,i}(t)$ for $\nu_1 = 2\pi/\tau$, $\tau = 10$, $\nu_i = i\, \nu_1$, $i = 2, 3$. The heavy and thin lines in the right panel of the figure are the solutions $x_1(t)$ and $x_2(t)$ for these two forcing functions. These solutions result from $I_{c,i}(t)$ and $I_{s,i}(t)$ by elementary calculations.

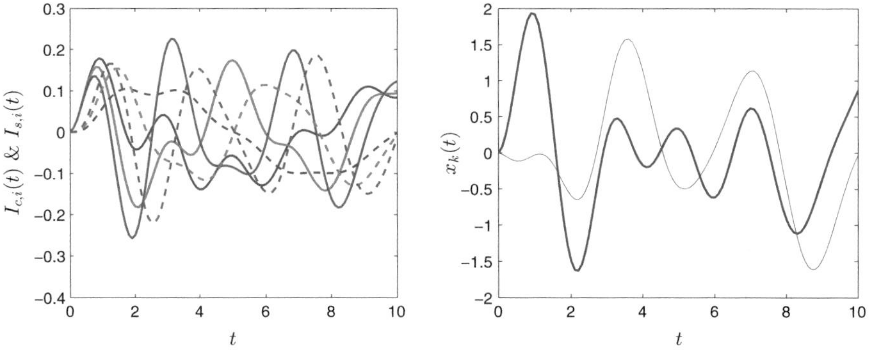

Fig. 2.16 Functions $I_{c,i}(t)$ and $I_{s,i}(t)$ (solid and dashed lines) and solutions $\{x_k(t)\}$ for two fording functions (left and right panels)

2.7 Applications

Our analysis can be applied directly to find the response of SDOF systems to seismic events via response spectra, find the response of simple systems to moving loads which may cause resonance, and calculate the displacement of SDOF systems with nonlinear damping. The following subsections illustrate these applications.

2.7.1 *Response Spectra*

Consider SDOF systems with parameters (ω, ζ) which is subjected to a seismic ground acceleration $a(t)$, $0 \le t \le \tau$, so that its displacement $x(t)$ is the solution of Eq. 2.32, i.e., $\ddot{x}(t) + 2\,\zeta\,\omega\,\dot{x}(t) + \omega^2\,x(t) = -a(t)$, $0 \le t \le \tau$, where τ is the duration of the seismic event.

Denote the displacement, velocity, and acceleration functions of an oscillator with parameters (ω, ζ) by $x(t; \omega, \zeta)$, $\dot{x}(t; \omega, \zeta)$, and $\ddot{x}(t; \omega, \zeta)$. The following response maxima

$$\begin{aligned} S_d(\omega, \zeta) &= \max_{0 \le t \le \tau} \left| x(t; \omega, \zeta) \right|, \\ S_v(\omega, \zeta) &= \max_{0 \le t \le \tau} \left| \dot{x}(t; \omega, \zeta) \right|, \\ S_a(\omega, \zeta) &= \max_{0 \le t \le \tau} \left| \ddot{x}(t; \omega, \zeta) + a(t) \right| \end{aligned} \tag{2.78}$$

are called the ***displacement***, ***velocity***, and ***absolute acceleration*** spectra. In Earthquake Engineering, the displacement spectrum $S_d(\omega, \zeta)$ is calculated while $S_v(\omega, \zeta)$ and $S_a(\omega, \zeta)$ are approximated by

$$\begin{aligned} S_v(\omega, \zeta) &\simeq P S_v(\omega, \zeta) = \omega\, S_d(\omega, \zeta) \\ S_a(\omega, \zeta) &\simeq P S_a(\omega, \zeta) = \omega^2\, S_d(\omega, \zeta) = \omega\, P S_v(\omega, \zeta). \end{aligned} \tag{2.79}$$

They are referred to as ***pseudo-velocity*** and ***pseudo-acceleration*** spectra. These equations provide the following two expressions of $P S_v$,

$$\begin{aligned} \log(P S_v) &= -\log(T) + \log(2\,\pi) + \log(S_d) \\ \log(P S_v) &= \log(T) - \log(2\,\pi) + \log(P S_a), \end{aligned} \tag{2.80}$$

where $T = 2\,\pi/\omega$ is the oscillator natural period. The two families of functions in Eq. 2.80 are lines at $\pm 45^o$ in the coordinates $\big(\log(T), \log(P S_v)\big)$ which are indexed by the damping ratio ζ. They are constructed for selected seismic ground accelerations $a(t)$ and are used to read the values of the displacement and pseudo-acceleration spectra for given period T and damping ratio ζ.

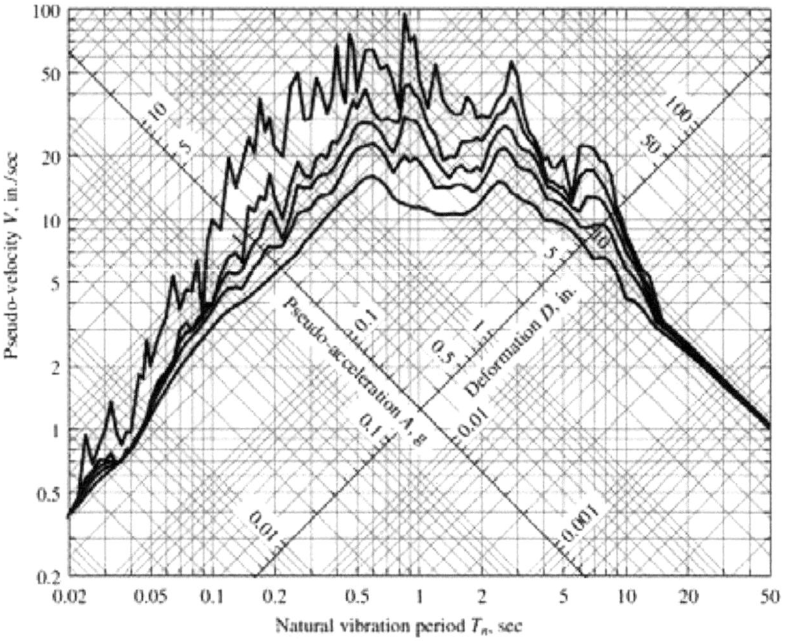

Fig. 2.17 El Centro response spectrum for $\zeta = 0, 2, 5$, and 10%. Note that the bottom and top spectral values are for $\zeta = 10\%$ and $\zeta = 0\%$

Example 2.17 Consider a portal frame with span and height $l = 24$ ft and $h = 12$ ft, square columns and beam with sides 10 in, and weight $w = 10$ kips. The frame is modeled as a SDOF system with stiffness $k = 96\,E\,I/(7\,h^3) = 11.48$ kips/in and natural period $T = 2\pi\sqrt{m/k} = 0.3$ s, where $E = 3 \times 10^3$ ksi is the modulus of elasticity, I denotes the moment of inertia of the beam and columns, and $m = w/g$ is the mass. Suppose the frame damping ratio is $\zeta = 5\%$.

The frame is subjected to the El Centro seismic ground acceleration. The maximum displacement of the frame and the maximum seismic force acting on it result directly from the response spectrum developed for this seismic event which is shown in Fig. 2.17. The maximum deformation of the frame during this seismic event can be read directly from this response spectrum; it is approximately 0.67 in. The maximum pseudo-acceleration can also be read from this figure; it is approximately 0.76 g. The corresponding seismic force can be obtained by multiplication with the frame mass. Note that the bottom and top spectral values are for $\zeta = 10\%$ and $\zeta = 0\%$.

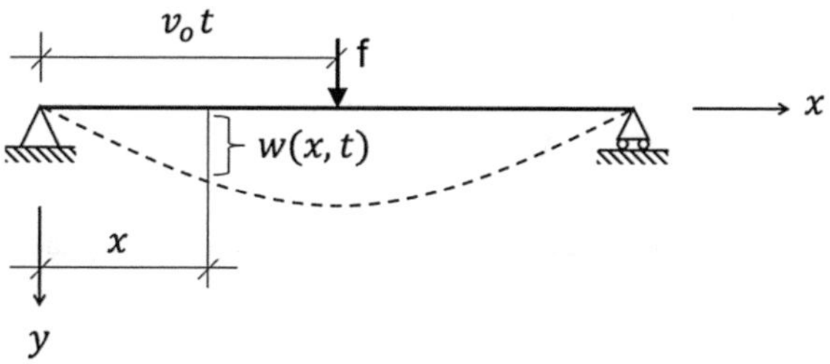

Fig. 2.18 Mathematical model of a single span bridge subjected to a concentrated moving load

2.7.2 *Moving Loads*

Consider a bridge modeled by the simply supported beam in Fig. 2.18 with span l and mass per unit length m. Assume that a vehicle (1) enters the bridge at time $t = 0$ and moves at a constant velocity v_0 so that its location at time t is $v_0\, t$ in the (x, y)-system of coordinates and (2) can be represented by a concentrated force f which is much smaller than the bridge weight, i.e., $f \ll m\, l\, g$, so that its contribution to dynamics can be disregarded.

The bridge has an infinite number of degree of freedom so that it cannot be analyzed directly by our current tools. However, our current tools suffice under the approximation that the bridge displacement $w(x, t)$, which is a function of the spatial coordinate x and time t, can be represented by

$$w(x, t) = \xi(t)\, \varphi(x), \quad 0 \le x \le l, \quad t \ge 0, \tag{2.81}$$

where $\varphi(x)$ is a specified function which satisfies the boundary conditions and $\xi(t)$ needs to be determined. This representation is used in the Chap. 5 to analyze continuous systems by the method of separation of variables for solving partial differential equations [3]. It also constitutes a simplified version of the Rayleigh-Ritz method [1] (Sect. 7.2).

The extended version of the work-energy formulation of Eq. 2.4 for rigid bodies which includes changes in the strain energy caused by deformation provides the tool for solving this problem. It can be written in the form $d\big(\mathrm{KE} + \mathrm{SE}\big) = d\,\mathrm{W}$, where KE, SE, and W denote the kinetic energy, strain energy, and external work [1] (Sect. 7.2). We now calculate these components of the work-energy equation and develop a differential equation for the unknown function $\xi(t)$.

– *Kinetic energy:* The kinetic energy of the beam element in $(x, x + dx)$ is

$$(1/2)\,(m\, dx)\, \dot{w}(x, t)^2 = (1/2)\,(m\, dx)\, \dot{\xi}(t)^2\, \varphi(x)^2$$

so that the beam kinetic energy is

$$\mathrm{KE} = \int_0^l (1/2)\,(m\, dx)\, \dot{w}(x, t)^2 = (m/2)\, \dot{\xi}(t)^2 \int_0^l \varphi(x)^2\, dx$$

and

$$d\ \text{KE} = \left(m \int_0^l \varphi(x)^2\, dx \right) \dot{\xi}(t)\, \ddot{\xi}(t)\, dt = m_{\text{eq}}\, \dot{\xi}(t)\, \ddot{\xi}(t)\, dt, \tag{2.82}$$

where m_{eq} has the meaning of an equivalent mass of the beam.

– *Strain energy:* It relates to deformation and has the expression [2]

$$\text{SE} = \frac{E\,I}{2} \int_0^l w''(x,t)^2\, dx = \frac{1}{2} \left(E\,I \int_0^l \left(\varphi''(x)\right)^2 dx \right) \xi(t)^2$$

so that

$$d\,\text{SE} = \left(E\,I \int_0^l \left(\varphi''(x)\right)^2 dx \right) \dot{\xi}(t)\, \xi(t)\, dt = k_{\text{eq}}\, \dot{\xi}(t)\, \xi(t)\, dt, \tag{2.83}$$

where E is the modulus of elasticity, I denotes the moment of inertia of the beam, and k_{eq} denotes the beam equivalent stiffness. Note that the equivalent mass and stiffness depend on the beam deformation.

– *External work:* The external work in a small time interval $(t, t + dt)$ is

$$f \left(\frac{\partial w(x,t)}{\partial x} (v_0\, dt) + \frac{\partial w(x,t)}{\partial t}\, dt \right) = f\, \left(\xi(t)\, \varphi'(x)\, v_0 + \dot{\xi}(t)\, \varphi(x) \right) dt$$
$$\simeq f\, \dot{\xi}(t)\, \varphi(x)\, dt \tag{2.84}$$

for $x = v_0\, t$, where the latter approximation holds under the assumption of small deformation such that $\varphi'(x) \ll \varphi(x)$.

These results and the work-energy equation $d\big(\text{KE} + \text{SE}\big) = d\,\text{W}$ give

$$m_{\text{eq}}\, \dot{\xi}(t)\, \ddot{\xi}(t)\, dt + k_{\text{eq}}\, \dot{\xi}(t)\, \xi(t)\, dt = f\, \dot{\xi}(t)\, \varphi(v_0\, t)\, dt$$

so that $\xi(t)$ is the solution of

$$m_{\text{eq}}\, \ddot{\xi}(t) + k_{\text{eq}}\, \xi(t) = f\, \varphi(v_0\, t), \quad t \geq 0, \tag{2.85}$$

which is the equation of motion for a SDOF with mass m_{eq}, stiffness k_{eq} and no damping which is subjected to the force $f\, \varphi(v_0\, t)$. Once $\xi(t)$ is calculated, the displacement of the beam can be obtained from Eq. 2.81.

Example 2.18 The function $\varphi(x) = \sin\left(\pi\, x/l\right)$ is valid since it satisfies all boundary conditions. The displacement function $w(x,t)$ is zero at supports, i.e., $w(0,t) = w(l,t) = 0$, at all times since $\varphi(0) = \varphi(l) = 0$ and so are the bending moments since $\varphi''(0) = \varphi''(l) = 0$. The equivalent mass and stiffness are

$$m_{eq} = m \int_0^l \sin^2(\pi x/l)\, dx = (m/2) \int_0^l (1 - \cos(2\pi x/l))\, dx = m\, l/2$$

$$k_{eq} = E\, I\, (\pi/l)^4 \int_0^l \sin^2(\pi x/l)\, dx = \frac{\pi^4 E\, I}{2\, l^3} \tag{2.86}$$

so that

$$\omega_{eq}^2 = \frac{k_{eq}}{m_{eq}} = \frac{\pi^4 E\, I}{m\, l^4}. \tag{2.87}$$

Then, $\xi(t)$ is the solution of

$$\ddot{\xi}(t) + \omega_{eq}^2\, \xi(t) = \frac{2\, f}{m\, l} \sin\big((\pi\, v_0/l)\, t\big), \tag{2.88}$$

which constitutes the equation of a SDOF system with no damping subjected to a harmonic force with frequency $\nu = \pi\, v_0/l$. If this frequency matches the natural frequency ω_{eq} of the system, i.e.,

$$v_0 = v_{0,cr} = \frac{\pi}{l} \sqrt{\frac{E\, I}{m}}, \tag{2.89}$$

the system is in resonance (see Eqs. 2.40 to 2.42). Its displacement increases indefinitely in time.

2.7.3 *Coulomb Damping Model*

Consider the SDOF system in Fig. 2.2 with no damper but friction between the system mass m and the supporting surface. Denote by μ the coefficient of friction. We only study the free vibration of this system. The x-coordinate has its origin at the undeformed position of the system and is positive if the spring is stretched. Consider an arbitrary time t. If $x(t) > 0$ and $\dot{x}(t) > 0$, the mass is in motion from left to right, so that the friction force is $-\mu\, m\, g$ as it opposes motion. The Newton law has the form $m\, \ddot{x} = -k\, x - \mu\, m\, g$. If $x(t) > 0$ and $\dot{x}(t) < 0$, the mass is in motion from right to left, so that the friction force is $\mu\, m\, g$. The Newton law gives $m\, \ddot{x} = -k\, x + \mu\, m\, g$. Accordingly, the equation of motion is

$$m\, \ddot{x} + \mu\, m\, g\, \text{sign}(\dot{x}) + k\, x = 0, \tag{2.90}$$

where $\text{sign}(\alpha) = 1$ and -1 for $\alpha > 0$ and $\alpha < 0$.

There is a significant difference between this equation of motion and that of Eq. 2.8 with $f(t) = 0$. The damping model of Eq. 2.90 is nonlinear in $\dot{x}$ while that of Eq. 2.8 is a linear function of $\dot{x}$. The methods of this chapter cannot be applied directly to solve Eq. 2.90 since the operator of this equation is nonlinear. However, we can use these methods to solve Eq. 2.49 by working on time intervals during which the motion is described by linear differential equations.

Suppose the initial conditions are $x_0 > 0$ and $\dot{x}_0 > 0$ so that the spring is stretched and the mass is in motion from left to right at the initial time. The equation of motion is

$$m\,\ddot{x} + \mu\,m\,g + k\,x = 0 \quad \text{or, equivalently, } \ddot{x} + \omega^2\,x = -\mu\,g, \quad (\omega^2 = k/m), \tag{2.91}$$

as long as the motion does not change direction. This is the equation of motion of a SDOF with no damping subjected to the force $-\mu\,g$. Its general solution has the expression

$$x(t) = A\,\cos(\omega\,t) + B\,\sin(\omega\,t) - \mu\,g/\omega^2. \tag{2.92}$$

The initial conditions give $A - \mu\,g/\omega^2 = x_0$ and $B\,\omega = \dot{x}_0$ so that

$$\begin{aligned} x(t) &= \left(x_0 + \frac{\mu\,g}{\omega^2}\right)\cos(\omega\,t) + \frac{\dot{x}_0}{\omega}\sin(\omega\,t) - \mu\,g/\omega^2 \\ \dot{x}(t) &= -\omega\left(x_0 + \frac{\mu\,g}{\omega^2}\right)\sin(\omega\,t) + \dot{x}_0\,\cos(\omega\,t). \end{aligned} \tag{2.93}$$

This solution is valid as long as the motion does not change direction, i.e., $\dot{x}(t) > 0$, so that $x(t)$ in Eq. 2.93 holds in the time interval $[0, t^*]$, where t^* is the solution of $\dot{x}(t^*) = 0$, i.e., $t^* = \tan^{-1}\left((\dot{x}_0/\omega)/(x_0 + \mu\,g/\omega^2)\right)$. The velocity changes sign at time t^* so that the mass moves from right to left for times $t > t^*$ as long as $\dot{x}(t)$ does not change sign.

Example 2.19 Suppose that the ICs are $x_0 > 0$ and $\dot{x}_0 = 0$. The mass does not move if the restoring force $k\,x_0$ of the spring is smaller than the friction force $\mu\,m\,g$. It moves from right to left if $k\,x_0 > \mu\,m\,g$ or $x_0 > \mu\,m\,g/k = \mu\,g/\omega^2$. Under this condition, the equation of motion is

$$\ddot{x} + \omega^2\,x = \mu\,g \tag{2.94}$$

so that

$$\begin{aligned} x(t) &= \left(x_0 - \frac{\mu\,g}{\omega^2}\right)\cos(\omega\,t) + \mu\,g/\omega^2 \\ \dot{x}(t) &= -\omega\left(x_0 - \frac{\mu\,g}{\omega^2}\right)\sin(\omega\,t) \end{aligned} \tag{2.95}$$

by arguments similar to those used to establish Eq. 2.93. Since $x_0 - \mu\, g/\omega^2 > 0$ by assumption, the velocity is negative so that the motion continues from right to left till time $t^* = \pi/\omega$ at which the velocity vanishes. The motion in the time interval $[0, t^*]$ is given by Eq. 2.95. It starts at x_0 and ends at $x(t^*) = -(x_0 - 2\,\mu\, g/\omega^2)$. The absolute value of the displacement at time t^* is smaller than x_0 since $|x(t^*)| = |(-x_0+\mu\, g/\omega^2)+\mu\, g/\omega^2| \leq \mu\, g/\omega^2 \leq x_0$ since $-x_0+\mu\, g/\omega^2 \leq 0$ by assumption and $\mu\, g/\omega^2 > 0$. This observation is consistent with the fact that friction dissipates energy. The next cycle can be treated in the same manner. The initial displacement and velocity for this cycle are $x(t^*) = -(x_0 - 2\,\mu\, g/\omega^2)$ and $\dot{x}(t^*) = 0$.

2.8 Problems

Problem 2.1 Repeat the calculations in Example 2.1 by using the coordinate y whose origin is at the top of the undeformed spring and is pointing upward.

Problem 2.2 The displacement $x(t)$ of a single degree of freedom system is the solution of

$$2\,\ddot{x}(t) + 1.6\pi\,\dot{x}(t) + 128\,x(t) = 10\,\cos(3t - 2.34), \quad t \geq 0.$$

Find the natural frequency ω and damping ratio ζ of the system.

Problem 2.3 Find the particular solution of the undamped SDOF system in Eq. 2.36 for the forcing function $f(t) = q\,\cos(\nu\,t)$, $\nu \neq \omega$.

Problem 2.4 Find the constants c_1 and c_2 in the expression of the homogeneous solution $x_h(t)$ in Eq. 2.19 from the initial conditions $(x_0, \dot{x}_0)$.

Problem 2.5 Complete the calculations following Eq. 2.46 to find the coefficients α and β in the expression of the particular solution $x_p(t)$.

Problem 2.6 Show that Eq. 2.45 is an alternative form of Eq. 2.44.
Hint: Write $x_p(t)$ of Eq. 2.44 in the form

$$x_p(t) = x_{\,\text{st}}\, r_d(\nu)\big[-2\,\zeta\,(\nu/\omega)\, r_d(\nu)\,\cos(\nu\,t) + \big((1-(\nu/\omega)^2\big)\, r_d(\nu)\,\sin(\nu\,t)\big],$$

use the notations $\sin(\varphi) = 2\,\zeta\,(\nu/\omega)\, r_d(\nu)$ and $\cos(\varphi) = \big((1-(\nu/\omega)^2\big)\, r_d(\nu)$, convince yourself that they are admissible, and complete the analysis by employing a trigonometric identity.

Problem 2.7 Repeat the calculations in Problem 2.3 for the forcing function $f(t) = q\,\cos(\nu\,t)$ and show that the particular solution has the expression $x_p(t) = x_{\text{st}}\, r_d(\nu)\,\cos(\nu\,t - \varphi)$ with the notations of Eqs. 2.44 and 2.45.

Problem 2.8 An undamped single degree of freedom system with natural frequency $\omega = 3\,\pi/2$ rad/sec is subjected to a seismic ground acceleration

$$a(t) = \sum_{k=1}^{n} \left[a_k \cos(\nu_k t) + b_k \sin(\nu_k t)\right], \quad 0 \le t \le \tau,$$

where $n = 100$, $\nu_k = 2\pi\, k/\tau$, $\tau = 10\,\mathrm{s}$, $a_k = \sin(\pi/k)$, and $b_k = \sin\left(3\pi/(2\,k)\right)$. Calculate and plot the Fourier transforms of the ground acceleration and the displacement $x(t)$ of the system. Compare your results with those of Example 2.14. Consider as in this example only the particular solution.

Problem 2.9 Repeat the calculation in the previous problem and assume that oscillator is damped with damping ratio $\zeta = 0.05$. Consider only the steady-state solution. Compare the Fourier transforms of the damped and undamped oscillators.

Problem 2.10 Consider an undamped SDOF system with stiffness $k = 10$ N/m and mass $m = 10$ kg which is subjected to the harmonic force

$$f(t) = \begin{cases} q\cos(\nu\, t), & 0 \le t \le t_1 \\ 0, & t_1 < t, \end{cases}$$

where $q = 10\,\mathrm{N}$, and $t_1 = 100\,\mathrm{s}$.

Determine and plot the motions of mass m for *(a)* $\nu = 0.4$ rad/sec, *(b)* $\nu = 0.999$ rad/sec, and *(c)* $\nu = 0.9$ rad/sec. Assume zero initial conditions. Comment on your results.

Problem 2.11 Using Duhamel's integral, determine the response of the system in Problem 2.10 subjected to the forcing function

$$f(t) = \begin{cases} q_1, & 0 \le t \le t_1 \\ q_2, & t_1 < t, \end{cases}$$

where $q_1 = 10\ N$, $q_2 = 15\ N$, $t_1 = 10\,\mathrm{s}$.

Problem 2.12 Determine the motion of an undamped SDOF system subjected to the forcing function $f(t) = q\left(\exp(-at) - \exp(-bt)\right)$, $a, b > 0$. The oscillator is at rest at the initial time. Plot the oscillator displacement for $a/\omega = 0.05,\ 0.1$, and 0.5. Assume $\omega = 1$, stiffness $k = 1$, $b = 0.1$, and $q = 1$.

Problem 2.13 Repeat the analysis of the previous problem and assume that the SDOF is damped with $\zeta = 0.05$.

Problem 2.14 Use the FD method to find the displacement of the SDOF system in Example 2.15 under the forcing function $f(t) = \sin(\nu\, t)$, where $\nu = \omega$.

Problem 2.15 Find the displacement function of the SDOF system in Problem 2.8 by the Fourier series method of Sect. 2.6.2.

References

1. T.M. Atanackovic, A. Guran, *Theory of Elasticity for Scientists and Engineers* (Birkhäuser, Boston, 1999)
2. E.P. Popov, *Engineering Mechanics of Solids* (Prentice Hall, New York, 1998)
3. D.A. Sánchez, *Ordinary Differential Equations and Stability Theory: An Introduction* (Dover Publications Inc., Mineola, 1968)
4. G.P. Tolstov, *Fourier Series* (Dover Publications Inc., New York, 1962)

Chapter 3
Eigenvalue Problem

Solutions of systems of linear algebraic equations are briefly reviewed and used to introduce the eigenvalue problem for square matrices. Properties of the eigenvalues and eigenvectors for symmetric real-valued matrices are first considered. These properties are then extended to real-valued nonsymmetric matrices.

The properties of the eigenvalues and eigenvectors of symmetric and nonsymmetric matrices in this chapter and the methods for analyzing SDOF systems in Chap. 2 are used to calculate the dynamical response of MDOF and continuous systems in Chaps. 4 and 5.

3.1 Systems of Linear Algebraic Equations

Consider the linear system of algebraic equations $\mathbf{a}\,\mathbf{x} = \mathbf{b}$, where $\mathbf{a}$ is an (n, n)-matrix and $\mathbf{b}$ is an $(n, 1)$-matrix, i.e., an n-dimensional column vector. It is assumed that the entries of the matrix $\mathbf{a}$ and the vector $\mathbf{b}$ are real numbers.

The system of equations is said to be ***inhomogeneous*** if the right side is not zero, i.e., $\mathbf{b} \neq \mathbf{0}$, and ***homogeneous*** otherwise, i.e., $\mathbf{b} = \mathbf{0}$. The solutions of these two types of systems differ significantly.

- *Inhomogeneous systems*: If the inverse $\mathbf{a}^{-1}$ of $\mathbf{a}$ exists, the system of equations has a unique solution given by $\mathbf{x} = \mathbf{a}^{-1}\,\mathbf{b}$.
- *Homogeneous systems*: The ***trivial*** solution, i.e., the null vector $\mathbf{x} = \mathbf{0}$, is always a solution. However, if $\det(\mathbf{a}) = 0$, the trivial solution is not the only solution. Under this condition, the homogeneous system of equations $\mathbf{a}\,\mathbf{x} = \mathbf{0}$ admits ***non-trivial*** solutions $\mathbf{x} \neq \mathbf{0}$ since its equations are not linearly independent, i.e., one or more equations are linear forms of the other equations.

M. D. Grigoriu, *Linear Dynamical Systems*,
https://doi.org/10.1007/978-3-030-64552-6_3

Example 3.1 The determinant of the (2, 2)-matrix

$$\mathbf{a} = \begin{bmatrix} 2 & 3 \\ 4 & 6 \end{bmatrix}$$

is $\det(\mathbf{a}) = (2)(6) - (3)(4) = 0$ so that we expect that the homogeneous system $\mathbf{a}\,\mathbf{x} = \mathbf{0}$ admits non-trivial solutions. To see this, set $x_1 = c \neq 0$, where c is an arbitrary non-zero constant. From the first equation, we have $x_2 = -2\,c/3$. Since $x_1 = c$ and $x_2 = -2\,c/3$ also satisfies the second equation, the system has the infinite set of non-trivial solutions $\mathbf{x}^T = [c \;\; -2\,c/3]$ indexed by c, which can be any non-zero real number. Note that the second equation $4\,x_1 + 6\,x_2 = 0$ of the homogeneous system $\mathbf{a}\,\mathbf{x} = \mathbf{0}$ is the first equation, $2\,x_1 + 3\,x_2 = 0$, multiplied by 2. The two equations of the system are linearly dependent (see Appendix C).

3.2 Eigenvalue Problem

Consider an (n, n)-matrix $\mathbf{a}$ and the system of linear algebraic equations $\mathbf{a}\,\mathbf{x} = \lambda\,\mathbf{x}$, where λ is a scalar. An alternative form of this system is $\left(\mathbf{a} - \lambda\,\mathbf{I}\right)\mathbf{x} = \mathbf{0}$, where $\mathbf{I}$ denotes the (n, n)-identity matrix, i.e., an (n, n)-diagonal matrix with unit non-zero entries. We have seen that this system of equations admits non-trivial solutions if the determinant of the matrix $\left(\mathbf{a} - \lambda\,\mathbf{I}\right)$ is zero. This condition is satisfied by the roots of the n-degree polynomial $\det\left(\mathbf{a} - \lambda\,\mathbf{I}\right)$. We limit the discussion to real-valued matrices $\mathbf{a}$, i.e., matrices $\mathbf{a}$ whose entries are reals.

Definition 3.1 The roots $\lambda_1, \ldots, \lambda_n$ of the n-degree polynomial $\det\left(\mathbf{a} - \lambda\,\mathbf{I}\right)$, i.e., the solutions of $\det\left(\mathbf{a} - \lambda\,\mathbf{I}\right) = 0$, are called the ***eigenvalues*** of $\mathbf{a}$. Since the determinant of a matrix and its transpose coincide, e.g., the determinants of matrices $\left(\mathbf{a} - \lambda\,\mathbf{I}\right)$ and $\left(\mathbf{a} - \lambda\,\mathbf{I}\right)^T = \left(\mathbf{a}^T - \lambda\,\mathbf{I}\right)$, the matrix $\mathbf{a}$ and its transpose $\mathbf{a}^T$ have the same eigenvalues. The eigenvalues $\lambda_1, \ldots, \lambda_n$ may or may not be distinct and can be real or complex.

Definition 3.2 Let λ_k be an eigenvalue of an (n, n) real-valued matrix $\mathbf{a}$ or, equivalently, of its transpose $\mathbf{a}^T$. An n-dimensional vector $\mathbf{x}_k$ is a ***right eigenvector*** of this matrix corresponding to λ_k if (1) it is not the null vector, i.e., $\mathbf{x}_k \neq \mathbf{0}$, and (2) it satisfies the linear system of equations $\mathbf{a}\,\mathbf{x}_k = \lambda_k\,\mathbf{x}_k$. An n-dimensional vector $\mathbf{x}_k$ is a ***left eigenvector*** of matrix $\mathbf{a}$ corresponding to λ_k if (1) it is not the null vector, i.e., $\mathbf{x}_k \neq \mathbf{0}$, and (2) it satisfies the linear system of equations $\mathbf{a}^T\,\mathbf{x}_k = \lambda_k\,\mathbf{x}_k$. If $\mathbf{a}$ is symmetric, the right and left eigenvectors coincide, and are referred to as ***eigenvectors*** of $\mathbf{a}$.

We list properties of the eigenvalues and eigenvectors of symmetric and nonsymmetric matrices which result directly from the above definitions. They are useful for numerical applications and analytical derivations.

1. The null vector $\mathbf{x} = \mathbf{0}$ is not a right/left eigenvector or eigenvector although it satisfies the eigenvalue equation $\mathbf{a}\,\mathbf{x} = \lambda\,\mathbf{x}$ or $\mathbf{a}^T\,\mathbf{x} = \lambda\,\mathbf{x}$. However, it violates the first requirement of the above definitions.
2. The right/left eigenvectors of nonsymmetric matrices and the eigenvectors of symmetric matrices can be calculated up to a multiplicative constant. For example, let $\mathbf{x}_k$ be a right eigenvector corresponding to the eigenvalue λ_k, so that $\mathbf{x}_k \neq \mathbf{0}$ and $\mathbf{a}\,\mathbf{x}_k = \lambda_k\,\mathbf{x}_k$. This system of equations multiplied by an arbitrary scalar β becomes $\mathbf{a}\,(\beta\,\mathbf{x}_k) = \lambda_k\,(\beta\,\mathbf{x}_k)$ so that $\beta\,\mathbf{x}_k$ is also a right eigenvector of $\mathbf{a}$ corresponding to the same eigenvalue λ_k.
3. We refer to $\mathbf{x}_k$ in the above definition as a right/left eigenvector or eigenvector of $\mathbf{a}$ since eigenvalues can have multiple eigenvectors. For example, the (2,2)-matrix

$$\mathbf{a} = \begin{bmatrix} 2 & 0 \\ 3 & 2 \end{bmatrix} \Longrightarrow \det \begin{bmatrix} 2-\lambda & 0 \\ 3 & 2-\lambda \end{bmatrix} = (\lambda - 2)^2 = 0$$

has a single eigenvalue ($\lambda_1 = \lambda_2 = 2$) but two right/left eigenvectors (see the subsequent subsection and Appendix D).
4. The eigenvalues of nonsymmetric matrices can be real- and/or complex-valued and so are the right/left eigenvectors.

Calculation of Eigenvalues/Eigenvectors The hand calculation of eigenvalues and eigenvectors is feasible for small matrices. The following two-step approach can be used.

- *Step 1. Eigenvalues:* Find the roots $\{\lambda_k\}$, $k = 1, \ldots, n$, of the polynomial $\det\big(\mathbf{a} - \lambda\,\mathbf{I}\big)$, i.e., the solutions of

$$\det\big(\mathbf{a} - \lambda\,\mathbf{I}\big) = \det\big(\mathbf{a}^T - \lambda\,\mathbf{I}\big) = 0. \tag{3.1}$$

The roots are the eigenvalues of matrices $\mathbf{a}$ and $\mathbf{a}^T$. They may or may not be distinct. They are real- and/or complex-valued depending on $\mathbf{a}$.
- *Step 2. Eigenvectors:* The non-trivial solutions $\{\mathbf{x}_k\}$, $k = 1, \ldots, n$, of the homogeneous systems of equations

$$\begin{aligned} &\big(\mathbf{a} - \lambda_k\,\mathbf{I}\big)\,\mathbf{x}_k = \mathbf{0} \Longleftrightarrow \mathbf{a}\,\mathbf{x}_k = \lambda_k\,\mathbf{x}_k \quad \text{(Right eigenvectors)} \\ &\big(\mathbf{a}^T - \lambda_k\,\mathbf{I}\big)\,\mathbf{x}_k = \mathbf{0} \Longleftrightarrow \mathbf{a}^T\,\mathbf{x}_k = \lambda_k\,\mathbf{x}_k \quad \text{(Left eigenvectors)} \end{aligned} \tag{3.2}$$

are the right and left eigenvectors of $\mathbf{a}$. As previously stated, these systems of equations admit non-trivial solutions since the determinants of $\big(\mathbf{a}-\lambda\,\mathbf{I}\big)$ and $\big(\mathbf{a}^T - \lambda\,\mathbf{I}\big)$ are zero for $\lambda = \lambda_k$, $k = 1, \ldots, n$. To avoid confusion, we will denote the right and left eigenvectors corresponding to an eigenvalue λ_k by $\mathbf{u}_k$ and $\mathbf{v}_k$. If $\mathbf{a}$ is symmetric, i.e., $\mathbf{a} = \mathbf{a}^T$, the above equations coincide and so do their solutions, i.e., the right and left eigenvectors. The non-trivial solutions of either equation in Eq. 3.2 give the eigenvectors of $\mathbf{a}$.

MATLAB Calculation of Eigenvalues/Eigenvectors Hand calculations using Eqs. 3.1 and 3.2 are feasible for small-dimensional matrices. It is preferable to find the eigenvalues and eigenvectors of **a** by MATLAB from

$$
\begin{aligned}
&[\mathbf{u}, \mathbf{d}] = \mathrm{eig}(\mathbf{a}) \quad \text{(symmetric matrices)} \\
&[\mathbf{u}, \mathbf{d}] = \mathrm{eig}(\mathbf{a}) \quad \text{and} \quad [\mathbf{v}, \mathbf{d}] = \mathrm{eig}(\mathbf{a}') \quad \text{(nonsymmetric matrices)},
\end{aligned} \tag{3.3}
$$

where **u**, **v**, and **d** are (n, n)-matrices and the notation $\mathbf{a}'$ indicates matrix transposition in MATLAB. The columns of **u** in the first call are the eigenvectors of symmetric matrices **a**. The columns of **u** and **v** in the second call are the right and left eigenvectors of nonsymmetric matrices **a**. The diagonal matrix **d** gives the eigenvalues of **a**, which coincide with those of $\mathbf{a}^T$. The columns of **u** and **v** are aligned with those of **d**, e.g., column r of **u** and **v** and column r of d correspond to the eigenvalue-eigenvector pair r.

Example 3.2 The outputs **u** and **d** of Eq. 3.3 for the (3, 3)-symmetric matrix

$$\mathbf{a} = \begin{bmatrix} 5 & 3 & 1 \\ 3 & 4 & 2 \\ 1 & 2 & 3 \end{bmatrix}$$

are

$$\mathbf{u} = \begin{bmatrix} -0.4226 & -0.5999 & 0.6793 \\ 0.7461 & 0.1953 & 0.6366 \\ -0.5145 & 0.7759 & 0.3651 \end{bmatrix} \quad \text{and} \quad \mathbf{d} = \begin{bmatrix} 0.9213 & 0 & 0 \\ 0 & 2.7302 & 0 \\ 0 & 0 & 8.3485 \end{bmatrix}.$$

Since the matrix is symmetric, the left and right eigenvectors coincide and are called just eigenvectors. Denote by $\lambda_1 = 0.9213$, $\lambda_2 = 2.7302$, and $\lambda_3 = 8.3485$ the eigenvalues of this matrix. The first, second, and third columns of the MATLAB output **u** are the eigenvectors $\mathbf{x}_1$, $\mathbf{x}_2$, and $\mathbf{x}_3$ associated with the eigenvalues λ_1, λ_2 and λ_3, which are in the first, second, and third columns of **d**.

Example 3.3 Consider the (2, 2)-nonsymmetric matrix

$$\mathbf{a} = \begin{bmatrix} 0 & 1 \\ -36 & -0.6 \end{bmatrix}.$$

The outputs of Eq. 3.3 for **a** are

$$\mathbf{u} = \begin{bmatrix} -0.0082 - 0.1642\,i & -0.0082 + 0.1642\,i \\ 0.9864 + 0.0000\,i & 0.9864 + 0.0000\,i \end{bmatrix}$$

and

$$\mathbf{d} = \begin{bmatrix} -0.3000 + 5.9925\,i & 0.0000 + 0.0000\,i \\ 0.0000 + 0.0000\,i & -0.3000 - 5.9925\,i \end{bmatrix},$$

where $i = \sqrt{-1}$ denotes the imaginary unit. The output $\mathbf{v}$ of Eq. 3.3

$$\mathbf{v} = \begin{bmatrix} -0.9864 + 0.0000\,i & -0.9864 + 0.0000\,i \\ -0.0082 + 0.1642\,i & -0.0082 - 0.164\,i \end{bmatrix}.$$

The matrices $\mathbf{u}$ and $\mathbf{v}$ give the right and left eigenvectors of $\mathbf{a}$. The eigenvalues of $\mathbf{a}$ are non-zero entries of the diagonal matrix $\mathbf{d}$.

The following subsections present properties of eigenvectors and eigenvalues for real-valued square matrices $\mathbf{a}$. Symmetric matrices, i.e., matrices with the property $\mathbf{a} = \mathbf{a}^T$, are first considered. Then, nonsymmetric matrices are discussed.

3.2.1 Symmetric Matrices

Suppose that the (n, n)-matrix $\mathbf{a}$ is real-valued and symmetric. These types of matrices are encountered frequently in the dynamic analysis of multi-degree of freedom systems. We state and prove properties of the eigenvalues and eigenvectors of real-valued symmetric matrices $\mathbf{a}$ which are relevant to our discussion.

1. **The eigenvalues are real-valued.**

   Proof Suppose that $\mathbf{x}_k$ is an eigenvector of matrix $\mathbf{a}$ corresponding to an eigenvalue λ_k of this matrix, so that λ_k and $\mathbf{x}_k$ satisfy the equation $\mathbf{a}\,\mathbf{x}_k = \lambda_k\,\mathbf{x}_k$. We have

$$\mathbf{a}\,\mathbf{x}_k = \lambda_k\,\mathbf{x}_k \Longrightarrow \left(\mathbf{a}\,\mathbf{x}_k\right)^{*T} = \left(\lambda_k\,\mathbf{x}_k\right)^{*T} \Longrightarrow \mathbf{x}_k^{*T}\,\mathbf{a} = \lambda_k^*\,\mathbf{x}_k^{*T}$$
$$\Longrightarrow \mathbf{x}_k^{*T}\,\mathbf{a}\,\mathbf{x}_k = \lambda_k^*\,\mathbf{x}_k^{*T}\,\mathbf{x}_k$$

   by taking the complex conjugate (symbol *) and the transposition (symbol ^{T}) of $\mathbf{a}\,\mathbf{x}_k = \lambda_k\,\mathbf{x}_k$ and using matrix operations, i.e., $(\mathbf{a}\,\mathbf{x}_k)^T = \mathbf{x}_k^T\,\mathbf{a}^T$ and $(\mathbf{a}\,\mathbf{x}_k)^* = \mathbf{a}^*\,\mathbf{x}_k^*$. Note that $\mathbf{a} = \mathbf{a}^*$ since the entries of $\mathbf{a}$ are reals and that $\mathbf{a} = \mathbf{a}^T$ since $\mathbf{a}$ is symmetric. The latter equality results by right multiplication with $\mathbf{x}_k$.

   Note also that left multiplication of the defining equation $\mathbf{a}\,\mathbf{x}_k = \lambda_k\,\mathbf{x}_k$ of λ_k and $\mathbf{x}_k$ by $\mathbf{x}_k^{*T}$ gives $\mathbf{x}_k^{*T}\,\mathbf{a}\,\mathbf{x}_k = \lambda_k\,\mathbf{x}_k^{*T}\,\mathbf{x}_k$. The comparison of this equation with the final result of the above derivations, i.e.,

$$\mathbf{x}_k^{*T}\,\mathbf{a}\,\mathbf{x}_k = \lambda_k\,\mathbf{x}_k^{*T}\,\mathbf{x}_k \quad \text{and} \quad \mathbf{x}_k^{*T}\,\mathbf{a}\,\mathbf{x}_k = \lambda_k^*\,\mathbf{x}_k^{*T}\,\mathbf{x}_k,$$

implies $\lambda_k\,\mathbf{x}_k^{*T}\,\mathbf{x}_k = \lambda_k^*\,\mathbf{x}_k^{*T}\,\mathbf{x}_k$ or $(\lambda_k - \lambda_k^*)\,\mathbf{x}_k^{*T}\,\mathbf{x}_k = \mathbf{0}$. Since $\mathbf{x}_k^{*T}\,\mathbf{x}_k \neq 0$ ($\mathbf{x}_k$ is eigenvector!), we have $\lambda_k = \lambda_k^*$ so that λ_k is real. Example 3.2 illustrates this property.

2. **The eigenvectors corresponding to distinct eigenvalues are orthogonal** in the sense that

$$\mathbf{x}_k^T\,\mathbf{x}_l = \delta_{kl} \quad \text{and} \quad \mathbf{x}_k^T\,\mathbf{a}\,\mathbf{x}_l = \lambda_k\,\delta_{kl} \tag{3.4}$$

provided that $\lambda_k \neq \lambda_l$, $k \neq l$, and the eigenvectors are scaled to have unit length, where $\delta_{kl} = 1$ for $k = l$ and $\delta_{kl} = 0$ for $k \neq l$. The matrix form of Eq. 3.4 is

$$\mathbf{x}^T\,\mathbf{a}\,\mathbf{x} = \mathrm{diag}\{\lambda_k\}, \tag{3.5}$$

where $\mathbf{x} = [\mathbf{x}_1\;\mathbf{x}_2 \cdots \mathbf{x}_n]$ is an (n, n)-matrix whose columns are the eigenvectors of $\mathbf{a}$ and $\mathrm{diag}\{\lambda_k\}$ denotes an (n, n)-diagonal matrix whose non-zero entries are the eigenvalues $\{\lambda_k\}$ of $\mathbf{a}$.

Proof The eigenvalues and the eigenvectors, $\{\lambda_k\}$ and $\{\mathbf{x}_k\}$, of $\mathbf{a}$ satisfy the equations $\mathbf{a}\,\mathbf{x}_k = \lambda_k\,\mathbf{x}_k$, $k = 1, \ldots, n$. Consider two distinct eigenvalues $\lambda_k \neq \lambda_l$, $k \neq l$. Simple manipulations of the defining equations of the eigenvalue-eigenvector pairs k and l give

$$\begin{aligned}
&\mathbf{a}\,\mathbf{x}_k = \lambda_k\,\mathbf{x}_k \Longrightarrow \mathbf{x}_k^T\,\mathbf{a} = \lambda_k\,\mathbf{x}_k^T \Longrightarrow \mathbf{x}_k^T\,\mathbf{a}\,\mathbf{x}_l = \lambda_k\,\mathbf{x}_k^T\,\mathbf{x}_l\\
&\mathbf{a}\,\mathbf{x}_l = \lambda_l\,\mathbf{x}_l \Longrightarrow \mathbf{x}_k^T\,\mathbf{a}\,\mathbf{x}_l = \lambda_l\,\mathbf{x}_k^T\,\mathbf{x}_l.
\end{aligned} \tag{3.6}$$

For example, $\mathbf{a}\,\mathbf{x}_k = \lambda_k\,\mathbf{x}_k$ becomes $\mathbf{x}_k^T\,\mathbf{a} = \lambda_k\,\mathbf{x}_k^T$ by transposition and $\mathbf{x}_k^T\,\mathbf{a}\,\mathbf{x}_l = \lambda_k\,\mathbf{x}_k^T\,\mathbf{x}_l$ by right multiplication with $\mathbf{x}_l$. The latter two equalities in Eq. 3.6, i.e., $\mathbf{x}_k^T\,\mathbf{a}\,\mathbf{x}_l = \lambda_k\,\mathbf{x}_k^T\,\mathbf{x}_l$ and $\mathbf{x}_k^T\,\mathbf{a}\,\mathbf{x}_l = \lambda_l\,\mathbf{x}_k^T\,\mathbf{x}_l$, have the same left sides so that their right sides must coincide, i.e., $\lambda_k\,\mathbf{x}_k^T\,\mathbf{x}_l = \lambda_l\,\mathbf{x}_k^T\,\mathbf{x}_l$ or, equivalently, $(\lambda_k - \lambda_l)\,\mathbf{x}_k^T\,\mathbf{x}_l = 0$. Since $\lambda_k \neq \lambda_l$ by assumption, we conclude $\mathbf{x}_k^T\,\mathbf{x}_l = 0$. The above equalities, e.g., $\mathbf{x}_k^T\,\mathbf{a}\,\mathbf{x}_l = \lambda_l\,\mathbf{x}_k^T\,\mathbf{x}_l$, also shows that $\mathbf{x}_k^T\,\mathbf{a}\,\mathbf{x}_l = 0$ for $k \neq l$ and $\mathbf{x}_k^T\,\mathbf{a}\,\mathbf{x}_k = \lambda_k\,\mathbf{x}_k^T\,\mathbf{x}_k = \lambda_k$. The eigenvalues and eigenvectors of matrix $\mathbf{a}$ in Example 3.2 provide an illustration of this property.

3. **If the eigenvalues are distinct, the eigenvectors define a basis of the n-dimensional Euclidian space $\mathbb{R}^n$.**

Proof The unit vectors $\mathbf{i} = (1, 0, 0)$, $\mathbf{j} = (0, 1, 0)$, and $\mathbf{k} = (0, 0, 1)$ of the physical 3-dimensional space $\mathbb{R}^3$ are orthogonal in the sense of the first equality of Eq. 3.4 and define a basis of this space. The vectors of $\mathbb{R}^3$ can be represented uniquely by their projections on **i**, **j**, and **k** (see Appendix C).

Similar arguments hold in higher dimensional spaces, i.e., the n-dimensional Euclidian spaces $\mathbb{R}^n$. Under the assumption that the eigenvalues are distinct, the orthogonality conditions $\mathbf{x}_k^T \, \mathbf{x}_l = \delta_{kl}$ hold for all pairs of eigenvectors so that the eigenvectors are linearly independent and define a basis for $\mathbb{R}^n$. Any element of this space, i.e., any n-dimensional vector, can be represented by its projections on the eigenvectors of matrix $\mathbf{a}$ (see Appendix C).

4. **A matrix a is positive definite if and only if its eigenvalues are positive and distinct.**

Proof Recall that $\mathbf{a}$ is positive definite if $\mathbf{x}^T \, \mathbf{a}\, \mathbf{x} > 0$ for any n-dimensional non-zero vector $\mathbf{x}$, i.e., $\mathbf{x} \neq \mathbf{0}$, and that the eigenvalues of $\mathbf{a}$ are real since this matrix is assumed to be symmetric with real entries.

Suppose first that $\mathbf{a}$ is positive definite. Then $\mathbf{x}_k^T \, \mathbf{a}\, \mathbf{x}_k > 0$, $k = 1, \ldots, n$, since $\mathbf{x}_k$ is an n-dimensional vector. We conclude that $\lambda_k > 0$ since $\mathbf{x}_k^T \, \mathbf{a}\, \mathbf{x}_k = \lambda_k$, see the second equality on Eq. 3.4 with $k = l$.

Suppose now that $\lambda_k > 0$ and consider an arbitrary n-dimensional vector. This vector admits the representation $\mathbf{x} = \sum_{k=1}^n \alpha_k \, \mathbf{x}_k$, where $\{\alpha_k\}$ denote the projections of $\mathbf{x}$ on the system of coordinates defined by the eigenvectors $\{\mathbf{x}_k\}$. This representation is valid by the previous property (see also Appendix C). Direct calculations give

$$\mathbf{x}^T \, \mathbf{a}\, \mathbf{x} = \left(\sum_{k=1}^n \alpha_k \, \mathbf{x}_k \right)^T \mathbf{a} \left(\sum_{l=1}^n \alpha_l \, \mathbf{x}_l \right) = \sum_{k,l=1}^n \alpha_k \, \alpha_l \, \mathbf{x}_k^T \, \mathbf{a}\, \mathbf{x}_l$$

$$= \sum_{k,l=1}^n \alpha_k \, \alpha_l \, \lambda_l \, \delta_{kl} = \sum_{k=1}^n \alpha_k^2 \, \lambda_k$$

so that $\mathbf{x}^T \, \mathbf{a}\, \mathbf{x} = \sum_{k=1}^n \alpha_k^2 \, \lambda_k > 0$ since $\lambda_k > 0$ by assumption.

This property provides a simple criterion for checking whether a symmetric matrix is positive definite. For example, we conclude that matrix $\mathbf{a}$ in Example 3.2 is positive definite since its eigenvalues are positive.

5. **An eigenvalue λ_1 of multiplicity $q \geq 2$ can be associated with q linearly independent generalized eigenvectors given by**

$$\begin{aligned}
\mathbf{a}\, \mathbf{x}_1 &= \lambda_1 \, \mathbf{x}_1 \\
\mathbf{a}\, \mathbf{x}_2 &= \lambda_1 \, \mathbf{x}_2 + \mathbf{x}_1 \\
&\vdots \\
\mathbf{a}\, \mathbf{x}_q &= \lambda_1 \, \mathbf{x}_q + \mathbf{x}_{q-1},
\end{aligned} \tag{3.7}$$

where $\mathbf{x}_1$ is the eigenvector of $\mathbf{a}$ corresponding to eigenvalue λ_1. The generalized eigenvectors $\mathbf{x}_r$, $r = 2, \ldots, q$, can be calculated recursively from the above equations.

Proof The proof of this statement can be found in, e.g., [1] (Sect. 6.3). This reference also shows that the eigenvectors and the generalized eigenvectors span the n-dimensional space (see also Appendix C).

We also note that if $\{\mathbf{x}_1, \ldots, \mathbf{x}_q\}$ are eigenvectors of an eigenvalue λ_1, linear forms of these eigenvectors are also eigenvectors of λ_1. Let $\alpha_1\,\mathbf{x}_1 + \cdots + \alpha_q\,\mathbf{x}_q$ be such a form, where $\{\alpha_j\}$, $j = 1, \ldots, q$, are arbitrary coefficients. We have

$$\begin{aligned}\mathbf{a}\left(\alpha_1\,\mathbf{x}_1 + \cdots + \alpha_q\,\mathbf{x}_q\right) &= \alpha_1\,\mathbf{a}\,\mathbf{x}_1 + \cdots + \alpha_q\,\mathbf{a}\,\mathbf{x}_q \\ &= \lambda_1\left(\alpha_1\,\mathbf{x}_1 + \cdots + \alpha_q\,\mathbf{x}_q\right)\end{aligned}$$

so that $\alpha_1\,\mathbf{x}_1 + \cdots + \alpha_q\,\mathbf{x}_q$ is an eigenvector of λ_1.

We do not expand further on properties of generalized eigenvectors to keep the presentation condensed and focused on typical applications. The above comments are intended to (1) recognize that multiple eigenvalues are encountered in applications and (2) provide a framework for the analysis of dynamical systems with multiple eigenvalues. These considerations also apply to the generalized eigenvectors of nonsymmetric matrices which are discussed in the subsequent section.

Example 3.4 The (2,2)-matrix

$$\mathbf{a} = \begin{bmatrix} 1 & 0 \\ 0 & 1 \end{bmatrix}$$

has the eigenvalues $\lambda_1 = \lambda_2 = 1$. The first eigenvector $\mathbf{x}_1$ is the solution of $\mathbf{a}\,\mathbf{x}_1 = \lambda_1\,\mathbf{x}_1$ or $\mathbf{x}_1 = \mathbf{x}_1$ so that the components of this vector are arbitrary constants constrained by the condition $\mathbf{x}_1 \neq \mathbf{0}$. The generalized eigenvector $\mathbf{x}_2$ is defined by $\mathbf{a}\,\mathbf{x}_2 = \lambda_1\,\mathbf{x}_2 + \mathbf{x}_1$ which gives $x_{2,1} = x_{2,1} + x_{1,1}$ and $x_{2,2} = x_{2,2} + x_{1,2}$. If $x_{1,1} \neq 0$, then $x_{2,1} = 0$ from the first equation. From the second equation, the solution $x_{2,1} = 0$, and the condition $\mathbf{x}_2 \neq \mathbf{0}$, we conclude that $x_{1,2}$ must be zero. In summary, the components of $\mathbf{x}_1$ and $\mathbf{x}_2$ are (arbitrary constant, 0) and (0, arbitrary constant). This result is consistent with our intuition since $\mathbf{a}$ is the identity matrix so that its eigenvectors are aligned with the system of coordinates in which this matrix is defined.

Example 3.5 The eigenvalues and eigenvectors of the (2,2)-matrix symmetric matrix

$$\mathbf{a} = \begin{bmatrix} 2 & 1 \\ 1 & 3 \end{bmatrix}$$

delivered by the MATLAB function $[\mathbf{u}, \mathbf{d}] = \mathrm{eig}(\mathbf{a})$ are

```
u =
    -0.8507    0.5257
     0.5257    0.8507

d =
     1.3820         0
          0    3.6180
```

so that $\lambda_1 = 1.3820$ $\lambda_2 = 3.6180$. The components of the corresponding eigenvectors $\mathbf{u}_1$ and $\mathbf{u}_2$ are $(-0.8507, 0.5257)$ and $(0.5257, 0.8507)$. The eigenvectors are lines passing through the origin of the system of coordinates of the matrix **a** which are orthogonal since $u_{1,1} * u_{2,1} + u_{1,2} * u_{2,2} = (-0.8507) * (0.5257) + (0.5257) * (0.8507) = 0$.

3.2.2 *Nonsymmetric Matrices*

Suppose now that the real-valued (n, n)-matrices **a** is not symmetric, i.e., $\mathbf{a} \neq \mathbf{a}^T$. We will deal with these types of matrices in our analysis of MDOF systems with non-proportional damping. We state and prove properties of the eigenvalues and eigenvectors of **a** and present numerical illustrations of these properties. As previously stated, we denote the right and left eigenvectors by **u** and **v**.

1. **The eigenvalues are real- and/or complex-valued.**

   Proof Instead of a proof, we provide two examples illustrating this property of nonsymmetric matrices. First, the eigenvalues of the (2,2)-nonsymmetric matrix **a** in Example 3.3 are complex-valued. Second, two eigenvalues of the (3,3)-matrix

$$\mathbf{a} = \begin{bmatrix} 0 & 1 & 0 \\ -36 & -0.6 & 0 \\ 0 & 0 & 1 \end{bmatrix}$$

   are complex-valued (the (2, 2)-matrix in Example 3.3) and the third is real-valued and equal to 1.

2. **The right and left eigenvectors corresponding to distinct eigenvalues are orthogonal** in the sense that

$$\mathbf{v}_l^T \, \mathbf{u}_k = 0, \quad \text{and } \mathbf{v}_l^T \, \mathbf{a} \, \mathbf{u}_k = 0, \quad k \neq l \quad \text{or in matrix form}$$
$$\mathbf{v}^T \, \mathbf{u} = \text{diag}\{\mathbf{v}_k^T \, \mathbf{u}_k\} \quad \text{and} \quad \mathbf{v}^T \, \mathbf{a} \, \mathbf{u} = \text{diag}\{\mathbf{v}_k^T \, \mathbf{a} \, \mathbf{u}_k\}, \tag{3.8}$$

   where $\lambda_k \neq \lambda_l$ for $k \neq l$, $\{\mathbf{u}_k\}$ and $\{\mathbf{v}_l\}$ are the right and left eigenvectors of **a** and the columns of the (n, n)-matrices $\mathbf{u} = [\mathbf{u}_1 \; \mathbf{u}_2 \ldots \mathbf{u}_n]$ and $\mathbf{v} = [\mathbf{v}_1 \; \mathbf{v}_2 \ldots \mathbf{v}_n]$

are the right and left eigenvectors of $\mathbf{a}$ which can be obtained by calling the MATLAB function in Eq. 3.3.

Proof The conditions of Eq. 3.2 with the notations $\mathbf{u}$ and $\mathbf{v}$ for the right and left eigenvectors become

$$\mathbf{a}\,\mathbf{u}_k = \lambda_k\,\mathbf{u}_k \quad \text{and} \quad \mathbf{a}^T\,\mathbf{v}_l = \lambda_l\,\mathbf{v}_l. \tag{3.9}$$

The first equation gives $\mathbf{v}_l^T\,\mathbf{a}\,\mathbf{u}_k = \lambda_k\,\mathbf{v}_l^T\,\mathbf{u}_k$ by left multiplication with $\mathbf{v}_l^T$. The transposed of the second equation becomes $\mathbf{v}_l^T\,\mathbf{a}\,\mathbf{u}_k = \lambda_l\,\mathbf{v}_l^T\,\mathbf{u}_k$ by right multiplication with $\mathbf{u}_k$. The left sides of the resulting two equations coincide so that their right sides must coincide, i.e., $\lambda_k\,\mathbf{v}_l^T\,\mathbf{u}_k = \lambda_l\,\mathbf{v}_l^T\,\mathbf{u}_k$ or $(\lambda_k - \lambda_l)\,\mathbf{v}_l^T\,\mathbf{u}_k = 0$. Since $\lambda_k \neq \lambda_l$ by assumption, we have $\mathbf{v}_l^T\,\mathbf{u}_k = 0$. The above equalities also give $\mathbf{v}_l^T\,\mathbf{a}\,\mathbf{u}_k = 0$, $k \neq l$. The matrix form of the orthogonality condition in Eq. 3.8 results by direct calculations. The reader is encouraged to perform these calculations since orthogonality is used extensively in the latter part of this book.

Note also that the equality $\lambda_k\,\mathbf{v}_l^T\,\mathbf{u}_k = \lambda_l\,\mathbf{v}_l^T\,\mathbf{u}_k$ above holds for any k and l. For $k = l$, we have

$$\lambda_k = \frac{\mathbf{v}_k^T\,\mathbf{a}\,\mathbf{u}_k}{\mathbf{v}_k^T\,\mathbf{u}_k}, \quad i = 1, \ldots, n, \tag{3.10}$$

which constitutes a useful relationship for the dynamic analysis of MDOF systems.

3. **If the eigenvalues are distinct, the right eigenvectors and the left eigenvectors define bases of $\mathbb{R}^n$.**

   Proof It is shown in Appendix C that the sets of right and left eigenvectors are linearly independent. The orthogonality property of Eq. 3.8 can be used to calculate the components of arbitrary vectors of $\mathbb{R}^n$. An arbitrary element $\mathbf{x}$ of $\mathbb{R}^n$ admits the representation $\mathbf{x} = \sum_{i=1}^n \langle \mathbf{x}, \mathbf{u}_i \rangle\,\mathbf{u}_i$, where $\langle \mathbf{x}, \mathbf{u}_i \rangle$ denoted the projection of $\mathbf{x}$ on $\mathbf{u}_i$ (see Appendix C). The vector $\mathbf{x}$ admits a similar representation in the basis defined by the left eigenvectors of $\mathbf{a}$.

4. **Nonsymmetric matrices can be positive definite.**

   Proof For example, the (2, 2)-nonsymmetric matrix

$$\mathbf{a} = \begin{bmatrix} 2 & 0 \\ 2 & 2 \end{bmatrix} \Longrightarrow \mathbf{x}^T\,\mathbf{a}\,\mathbf{x} = (x_1 + x_2)^2 + x_1^2 + x_2^2 > 0$$

   is positive definite. We do not explore further properties of these types of matrices since they are not used in our discussion. The interest reader can consult [1] (Sect. 5.3).

5. **An eigenvalue λ_1 of multiplicity $q \geq 2$ can be associated with q linearly independent generalized right/left eigenvectors.**

   Proof See [1] (Sect. 6.3) for proof and technical considerations. The generalized right eigenvectors can be obtained by the algorithm of Eq. 3.7. The generalized left eigenvectors result from this equation with $\mathbf{a}^T$ in place of $\mathbf{a}$.

Example 3.6 The (2,2)-nonsymmetric matrix

$$\mathbf{a} = \begin{bmatrix} 1 & 1 \\ 0 & 1 \end{bmatrix}$$

has the eigenvalues $\lambda_1 = \lambda_2 = 1$. The first right eigenvector $\mathbf{x}_1$ given by MATLAB has the components $(1, 0)$. The second right eigenvector results from Eq. 3.7 with $q = 2$, i.e., $\mathbf{a}\,\mathbf{u}_2 = \lambda\,\mathbf{u}_2 + \mathbf{u}_1$ which gives the equations $u_{2,1} + u_{2,2} = u_{2,1} + 1$ and $u_{2,2} = u_{2,2}$ so that $u_{2,2} = 1$ and $u_{2,1}$ is an arbitrary constant, where $u_{k,r}$ denotes the rth component of $\mathbf{u}_k$.

3.3 Problems

Problem 3.1 Construct examples of linear homogeneous systems of algebraic equations with $n \geq 3$ which admit non-trivial solutions and find their solutions.

Problem 3.2 Consider an (n, n)-symmetric real-valued matrix $\mathbf{a}$ with eigenvectors $\{\mathbf{x}_i\}$ assumed to have unit length. Show that the (n, n)-matrix $\mathbf{X}^T\,\mathbf{a}\,\mathbf{X}$ is diagonal with non-zero entries $\{\lambda_k\}$, where $\mathbf{X} = [\mathbf{x}_1\ \mathbf{x}_2 \ldots \mathbf{x}_n]$.

Hint: Use the orthogonality conditions of Eq. 3.4.

Problem 3.3 Find the eigenvalues and the eigenvectors of the $(2, 2)$-matrix

$$\mathbf{a} = \begin{bmatrix} 10 & -3 \\ -3 & 4 \end{bmatrix}$$

by hand calculations and by MATLAB. Compare results and plot the eigenvectors.

Problem 3.4 Find by hand calculations the eigenvalues and the right/left eigenvectors of the $(2, 2)$-matrix

$$\mathbf{a} = \begin{bmatrix} 2 & 1 \\ 0 & 1 \end{bmatrix}.$$

Show that they are orthogonal in the sense of Eq. 3.8.

Problem 3.5 Find the eigenvalues and the eigenvectors of the (3, 3)-matrix

$$\mathbf{a} = \begin{bmatrix} 1 & 1 & 2 \\ 1 & 1 & 3 \\ 2 & 3 & 2 \end{bmatrix},$$

plot the eigenvectors and show that they are orthogonal in the sense of Eq. 3.4.

Problem 3.6 Find the eigenvalues and the right/left eigenvectors of the (2, 2)-matrix

$$\mathbf{a} = \begin{bmatrix} 1 & 5 \\ 0 & 3 \end{bmatrix}$$

by hand calculations and MATLAB. Show that the eigenvectors are orthogonal sense of Eq. 3.4.

Problem 3.7 Consider a vector $\mathbf{x}$ in $\mathbb{R}^3$ with components $(1, -2, 3)$ in the standard reference $\{\mathbf{i}_k\}$, $k = 1, 2, 3$. Find the components of $\mathbf{x}$ in the coordinates defined by the eigenvectors $\{\mathbf{x}_k\}$, $k = 1, 2, 3$, of the matrix in Problem 3.5. Plot the eigenvectors $\{\mathbf{x}_k\}$ and the vector $\mathbf{x}$ in the reference $\{\mathbf{i}_k\}$. Plot also the unit vectors $\{\mathbf{i}_k\}$ and the vector $\mathbf{x}$ in the reference defined by the eigenvectors $\{\mathbf{x}_k\}$.

Reference

1. P. Lancaster, M. Tismenetsky, *The Theory of Matrices*, 2nd edn. (Academic Press Inc., New York, 1985)

Chapter 4
Multi-Degree of Freedom (MDOF) Systems

We consider systems with finite numbers $n > 1$ of degrees of freedom. Systems with infinite numbers of degrees of freedom, referred to as continuous systems, are discussed in the subsequent chapter. It will be seen that the methods for solving MDOF and continuous systems are conceptually similar and involve three steps.

First, the displacement vectors/functions of MDOF/continuous systems are viewed as elements of the linear spaces spanned by the eigenvectors/eigenfunctions of these systems. The methods of Chap. 3 are employed to construct eigenvector/eigenfunction coordinates. The displacement functions are completely defined by their projections on these coordinates, which are finite for MDOF systems and (countable) infinite for continuous systems.

Second, differential equations are developed for the projections of the displacement vectors/functions of MDOF/continuous systems on their eigenvectors/eigenfunctions. These equations are uncoupled and have the structure of the equations of motion for SDOF systems. They can be solved one-by-one by using the methods developed in Chap. 2.

Third, the displacement vectors/functions of MDOF/continuous systems are assembled from their representations in the eigenvector/eigenfunction coordinates and their projections of these coordinates delivered by the previous step.

In summary, the analysis of MDOF and continuous systems does not introduce new concepts. It uses tools for solving equations of motion for SDOF systems and properties of eigenvectors and eigenvalues, which are discussed in the previous two chapters.

M. D. Grigoriu, *Linear Dynamical Systems*,
https://doi.org/10.1007/978-3-030-64552-6_4

4.1 Physical System

Consider the 2-DOF system in Fig. 4.1 whose masses m_1 and m_2 are connected by springs with stiffnesses k_1 and k_2 and dampers with coefficients c_1 and c_2. The system is subjected to the forcing function $\mathbf{f}(t)$ with components $f_1(t)$ and $f_2(t)$ that act on the system masses. The figure also shows the free-body diagram of this system in the (x_1, x_2)-coordinates with origin corresponding to the undeformed springs.

4.2 Equations of Motion

The Newton law applied to the masses m_1 and m_2 of the system in Fig. 4.1 gives

$$\begin{aligned} m_1\,\ddot{x}_1 &= -c_1\,\dot{x}_1 + c_2\,(\dot{x}_2 - \dot{x}_1) - k_1\,x_1 + k_2\,(x_2 - x_1) + f_1(t) \\ m_2\,\ddot{x}_1 &= -c_2\,\dot{x}_2 - k_2\,(x_2 - x_1) + f_2(t), \end{aligned} \tag{4.1}$$

or, in matrix form,

$$\mathbf{m}\,\ddot{\mathbf{x}}(t) + \mathbf{c}\,\dot{\mathbf{x}}(t) + \mathbf{k}\,\mathbf{x}(t) = \mathbf{f}(t), \tag{4.2}$$

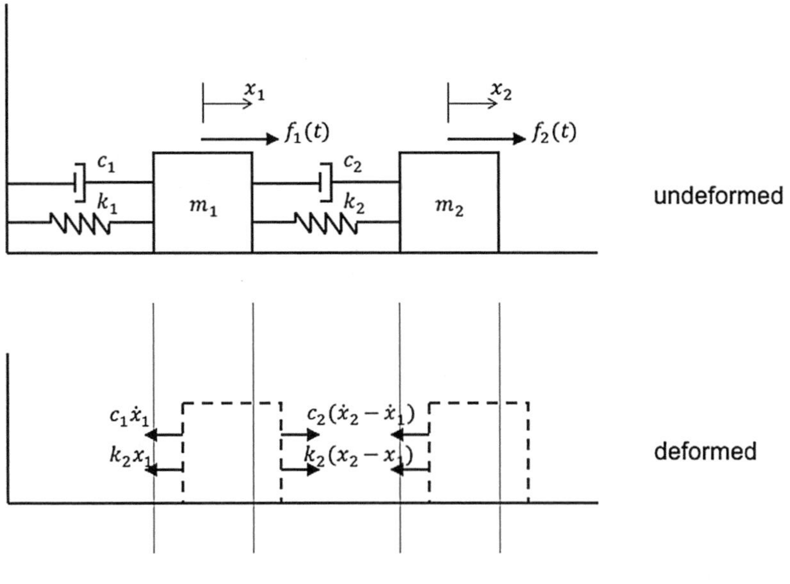

Fig. 4.1 Physical model for 2-degree of freedom systems and free-body diagram

where

$$\mathbf{m} = \begin{bmatrix} m_1 & 0 \\ 0 & m_2 \end{bmatrix}, \quad \mathbf{c} = \begin{bmatrix} c_1 + c_2 & -c_2 \\ -c_2 & c_2 \end{bmatrix}, \quad \mathbf{k} = \begin{bmatrix} k_1 + k_2 & -k_2 \\ -k_2 & k_2 \end{bmatrix},$$

$$\mathbf{x}(t) = \begin{bmatrix} x_1(t) \\ x_2(t) \end{bmatrix} \quad \text{and} \quad \mathbf{f}(t) = \begin{bmatrix} f_1(t) \\ f_2(t) \end{bmatrix} \tag{4.3}$$

denote the mass, damping, and stiffness matrices and the displacement vector with components $x_1(t)$ and $x_2(t)$. These considerations extend directly to n-DOF systems in which case $\mathbf{m}$, $\mathbf{c}$, and $\mathbf{k}$ are (n, n)-matrices and $\mathbf{x}(t)$ and $\mathbf{f}(t)$ are $(n, 1)$-matrices at each time, i.e., n-dimensional column vectors.

Our objective is to find the solution of Eq. 4.2 for specified initial displacement $\mathbf{x}_0$ and velocity $\dot{\mathbf{x}}_0$, i.e., the displacement vector $\mathbf{x}(t)$ of MDOF systems subjected to specified forcing functions and initial conditions. In addition to the solution of Eq. 4.2 that constitutes the ***forced vibration*** of damped MDOF systems, we also consider special cases of this equation. For example, the solution of the homogeneous version of Eq. 4.2, i.e., this equation with $\mathbf{f}(t) = \mathbf{0}$, constitutes the ***free vibration*** of damped MDOF systems caused by the initial conditions $(\mathbf{x}_0, \dot{\mathbf{x}}_0)$. The solutions of Eq. 4.2 with $\mathbf{c} = \mathbf{0}$ constitute the ***free*** or ***forced vibration*** of undamped MDOF systems for $\mathbf{f}(t) = \mathbf{0}$ or $\mathbf{f}(t) \neq \mathbf{0}$.

We construct solutions of Eq. 4.2 and special cases of this equation by analysis in the time and frequency domains. The particulars of these solutions depend on the properties of the damping matrix.

4.3 Classical Modal Analysis

We have seen that the free vibration solution of undamped SDOF systems has the form $\sin(\omega t + \varphi)$ (see Eq. 2.23 with $\zeta = 0$). A similar form works for the solution of Eq. 4.2 with $\mathbf{c} = \mathbf{0}$ and $\mathbf{f}(t) = \mathbf{0}$, i.e., $\mathbf{x}(t) = \mathbf{\Phi} \sin(\omega t + \varphi)$, where $\mathbf{\Phi}$ is an n-dimensional column vector. Rigorous consideration of the form of the free vibration solution of undamped MDOF systems can be found in, e.g., [2, Sect. 6.3], [4, Chap. 22], and [1, Chap. 1].

The solution $\mathbf{x}(t) = \mathbf{\Phi} \sin(\omega t + \varphi)$ and Eq. 4.2 with $\mathbf{c} = \mathbf{0}$ and $\mathbf{f}(t) = \mathbf{0}$ imply the condition

$$\left(-\mathbf{m}\,\omega^2 + \mathbf{k}\right) \mathbf{\Phi} \sin(\omega t + \varphi) = \mathbf{0},$$

which has to be satisfied at all times $t \geq 0$. Since $\sin(\omega t + \varphi)$ cannot be zero at all times, we must require

$$\left(-\mathbf{m}\,\omega^2 + \mathbf{k}\right) \mathbf{\Phi} = \mathbf{0}. \tag{4.4}$$

This condition is a homogeneous system of n linear algebraic equations for $\mathbf{\Phi}$. Its trivial solution $\mathbf{\Phi} = \mathbf{0}$ is not acceptable for non-zero initial conditions since, if $\mathbf{\Phi} = \mathbf{0}$, then $\mathbf{x}(t) = \mathbf{0}$ at all times, which would be at variance with non-zero initial conditions. Our interest is in non-trivial solutions of this algebraic system of equations.

4.3.1 Modal Frequencies and Shapes

We have shown that a homogeneous system of equations admits non-trivial solutions if its determinant is zero. Accordingly, the homogeneous system in Eq. 4.4 admits non-trivial solutions $\mathbf{\Phi} \neq \mathbf{0}$ if the determinant of matrix $\left(-\mathbf{m}\,\omega^2+\mathbf{k}\right)$ is zero. Since this matrix depends on the unspecified parameter ω^2, it is likely that the requirement $\det\left(-\mathbf{m}\,\omega^2+\mathbf{k}\right) = 0$ can be satisfied. It turns out that the values of ω^2 for which $\det\left(-\mathbf{m}\,\omega^2+\mathbf{k}\right) = 0$ have physical meaning and so do the corresponding non-trivial solutions of Eq. 4.4.

The solution of the homogeneous system of algebraic equations in Eq. 4.4 involves the following two steps.

– *Step 1.* ***Modal frequencies:*** Denote by $\omega_1^2, \ldots, \omega_n^2$ the roots of the n-degree polynomial $\det\left(-\mathbf{m}\,\omega^2+\mathbf{k}\right)$ in ω^2, which are given by the solutions of

$$\det\left(-\mathbf{m}\,\omega^2+\mathbf{k}\right) = 0. \tag{4.5}$$

The modal frequencies are the square roots $\omega_1, \ldots, \omega_n$ of the solutions $\omega_1^2, \ldots, \omega_n^2$. They are indexed such that $\omega_1 \leq \omega_2 \leq \cdots \leq \omega_n$. If the roots are distinct, the inequalities are strict, i.e., we have $\omega_1 < \omega_2 < \cdots < \omega_n$. Note that:

1. The modal frequencies can also be obtained from

$$\det\left(\mathbf{m}^{-1}\,\mathbf{k} - \omega^2\,\mathbf{I}\right) = 0, \tag{4.6}$$

since $\det\left(-\mathbf{m}\,\omega^2+\mathbf{k}\right) = \det(\mathbf{m})\,\det\left(-\mathbf{I}\,\omega^2+\mathbf{m}^{-1}\,\mathbf{k}\right)$ and $\det(\mathbf{m}) \neq 0$. The matrix $\mathbf{m}^{-1}\,\mathbf{k}$ is the analogue of the matrix $\mathbf{a}$ in our discussion on the eigenvalue problem (see Eq. 3.1).
2. The notation ω^2 for the eigenvalues of $\mathbf{m}^{-1}\,\mathbf{k}$ is meaningful if $\mathbf{m}^{-1}\,\mathbf{k}$ is a real-valued, symmetric, and positive definite matrix since the eigenvalues of these types of matrices are real and positive (see properties of eigenvalues in Sect. 3.2.1). We note that $\mathbf{m}^{-1}\,\mathbf{k}$ is a real-valued, symmetric, and positive definite matrix for the problems considered in the book.

– *Step 2.* ***Modal shapes:*** The non-trivial solutions $\{\mathbf{\Phi}_i\}$ of the homogeneous system of equations

$$\big(-\mathbf{m}\,\omega_i^2+\mathbf{k}\big)\,\mathbf{\Phi}_i=\mathbf{0}\quad \text{or, equivalently,}\quad \mathbf{k}\,\mathbf{\Phi}_i=\omega_i^2\,\mathbf{m}\,\mathbf{\Phi}_i, \tag{4.7}$$

i.e., the non-trivial solutions of Eq. 4.4 with $\{\omega^2=\omega_i^2\}$ are the modal shapes associated with the modal frequencies $\{\omega_i\}$, $i=1,\ldots,n$. Note that modal frequencies and shapes are system properties. They are completely defined by the system mass and stiffness matrices.

The modal shapes and frequencies of a MDOF with mass and stiffness matrices **m** and **k** are given by the MATLAB function (see Eqs. 3.3 and 4.7)

$$[\mathbf{u},\mathbf{d}]=\mathrm{eig}\big(\mathbf{m}^{-1}\,\mathbf{k}\big), \tag{4.8}$$

where the columns of the (n,n)-matrix **u** are the modal shapes, and the non-zero entries of the (n,n)-diagonal matrix **d** are the squares of the modal frequencies. The columns of **u** are paired with those of **d**, e.g., column r of **u**, i.e., the rth modal shape (eigenvector), corresponds to the entry (r,r) of **d**, which gives the rth modal frequency (eigenvalue).

4.3.2 Properties of Modal Shapes and Frequencies

The modal shapes are eigenvectors of real-valued, symmetric, and positive definite matrix $\mathbf{m}^{-1}\,\mathbf{k}$, and their properties are discussed in Sect. 3.2.1. We only discussed here the orthogonality of modal shapes.

1. **The modal shapes of distinct modal frequencies are orthogonal** in the sense that

$$\mathbf{\Phi}_i^T\,\mathbf{k}\,\mathbf{\Phi}_j=0\quad\text{and}\ \mathbf{\Phi}_i^T\,\mathbf{m}\,\mathbf{\Phi}_j=0,\, i\neq j$$

$$\mathbf{\Phi}_i^T\,\mathbf{k}\,\mathbf{\Phi}_i=\omega_i^2\,\mathbf{\Phi}_i^T\,\mathbf{m}\,\mathbf{\Phi}_i\Longrightarrow\omega_i^2=\frac{\mathbf{\Phi}_i^T\,\mathbf{k}\,\mathbf{\Phi}_i}{\mathbf{\Phi}_i^T\,\mathbf{m}\,\mathbf{\Phi}_i}=\frac{\tilde{k}_i}{\tilde{m}_i}, \tag{4.9}$$

where $\tilde{k}_i=\mathbf{\Phi}_i^T\,\mathbf{k}\,\mathbf{\Phi}_i$ and $\tilde{m}_i=\mathbf{\Phi}_i^T\,\mathbf{m}\,\mathbf{\Phi}_i$ denote the ***modal stiffness*** and ***modal mass*** of mode i.

Proof The definition $\mathbf{k}\,\mathbf{\Phi}_i=\omega_i^2\,\mathbf{m}\,\mathbf{\Phi}_i$ of modal shape i gives $\mathbf{\Phi}_j^T\,\mathbf{k}\,\mathbf{\Phi}_i=\omega_i^2\,\mathbf{\Phi}_j^T\,\mathbf{m}\,\mathbf{\Phi}_i$ by left multiplication with $\mathbf{\Phi}_j^T$.
Similarly, the definition $\mathbf{k}\,\mathbf{\Phi}_j=\omega_j^2\,\mathbf{m}\,\mathbf{\Phi}_j$ of mode j gives

$$\mathbf{\Phi}_j^T\,\mathbf{k}=\omega_j^2\,\mathbf{\Phi}_j^T\,\mathbf{m}\Longrightarrow\mathbf{\Phi}_j^T\,\mathbf{k}\,\mathbf{\Phi}_i=\omega_j^2\,\mathbf{\Phi}_j^T\,\mathbf{m}\,\mathbf{\Phi}_i$$

by transposition and right multiplication with $\mathbf{\Phi}_i$. Since the left sides of the above equations coincide, we have the equality of the corresponding right sides, i.e., $\omega_i^2\,\mathbf{\Phi}_j^T\,\mathbf{m}\,\mathbf{\Phi}_i = \omega_j^2\,\mathbf{\Phi}_j^T\,\mathbf{m}\,\mathbf{\Phi}_i$ or $\left(\omega_i^2 - \omega_j^2\right)\mathbf{\Phi}_j^T\,\mathbf{m}\,\mathbf{\Phi}_i = 0$, which implies $\mathbf{\Phi}_j^T\,\mathbf{m}\,\mathbf{\Phi}_i = 0$ since $\omega_i \neq \omega_j$ by assumption. The above equations also show that $\mathbf{\Phi}_j^T\,\mathbf{k}\,\mathbf{\Phi}_i = 0$, $i \neq j$, and that $\mathbf{\Phi}_i^T\,\mathbf{k}\,\mathbf{\Phi}_i = \omega_i^2\,\mathbf{\Phi}_i^T\,\mathbf{m}\,\mathbf{\Phi}_i$.

2. **The matrix form of orthogonality conditions** in Eq. 4.9 is

$$\mathbf{\Psi}^T\,\mathbf{m}\,\mathbf{\Psi} = \mathrm{diag}\{\tilde{m}_i\} \quad \text{and} \quad \mathbf{\Psi}^T\,\mathbf{k}\,\mathbf{\Psi} = \mathrm{diag}\{\tilde{k}_i\}, \tag{4.10}$$

where the columns of the (n, n)-matrix $\mathbf{\Psi} = [\mathbf{\Phi}_1\mathbf{\Phi}_2 \cdots \mathbf{\Phi}_n]$ are the system modal shapes. Note that the MATLAB output $\mathbf{u}$ in Eq. 4.8 is the matrix $\mathbf{\Psi}$.

4.3.3 Approximate Calculation of Modal Frequencies

Consider an n-DOF structure with modal frequencies $\omega_1 < \omega_2 < \cdots < \omega_n$ and modal shapes $\mathbf{\Phi_1}, \mathbf{\Phi_2}, \ldots, \mathbf{\Phi_n}$. We have established the following relationship (see Eq. 4.9):

$$\omega_i^2 = \frac{\mathbf{\Phi}_i^T\,\mathbf{k}\,\mathbf{\Phi}_i}{\mathbf{\Phi}_i^T\,\mathbf{m}\,\mathbf{\Phi}_i} = \frac{\tilde{k}_i}{\tilde{m}_i} \tag{4.11}$$

between modal shapes, modal frequencies, and mass/stiffness matrices. This relationship can be used to estimate modal frequencies from guesses of modal shapes based on the following two properties.

1. If $\mathbf{u}$ is a possible deformed shape, i.e., it satisfies geometrical boundary conditions, then

$$\omega_1^2 \leq \frac{\mathbf{u}^T\,\mathbf{k}\,\mathbf{u}}{\mathbf{u}^T\,\mathbf{m}\,\mathbf{u}} \leq \omega_n^2. \tag{4.12}$$

2. If $\mathbf{u} = \mathbf{\Phi}_i + \varepsilon\,\mathbf{v}$ is a possible deformed shape and ε is a small parameter, then

$$\frac{\mathbf{u}^T\,\mathbf{k}\,\mathbf{u}}{\mathbf{u}^T\,\mathbf{m}\,\mathbf{u}} = \omega_i^2 + O(\varepsilon). \tag{4.13}$$

Proof Since $\mathbf{u}$ is a possible deformation, it is a vector in the linear space spanned by modal shapes, i.e., $\mathbf{u} = \sum_{i=1}^n b_i\,\mathbf{\Phi}_i$, where $\{b_i\}$ are coefficients that determine the precise form of $\mathbf{u}$ (see Appendix C). This representation of $\mathbf{u}$ inserted in Eq. 4.11 gives

$$\omega^2 = \frac{\mathbf{u}^T\,\mathbf{k}\,\mathbf{u}}{\mathbf{u}^T\,\mathbf{m}\,\mathbf{u}} = \frac{\sum_{i,j=1}^n b_i\,b_j\,\mathbf{\Phi}_i^T\,\mathbf{k}\,\mathbf{\Phi}_j}{\sum_{i,j=1}^n b_i\,b_j\,\mathbf{\Phi}_i^T\,\mathbf{m}\,\mathbf{\Phi}_j} = \frac{\sum_{i=1}^n b_i^2\,\tilde{k}_i}{\sum_{i=1}^n b_i^2\,\tilde{m}_i}$$

(by orthogonality and Eq. 4.11)

$$= \frac{\sum_{i=1}^n b_i^2\,\omega_i^2\,\tilde{m}_i}{\sum_{i=1}^n b_i^2\,\tilde{m}_i} = \sum_{i=1}^n \frac{b_i^2\,\tilde{m}_i}{\sum_{j=1}^n b_j^2\,\tilde{m}_j}\,\omega_i^2 = \sum_{i=1}^n d_i^2\,\omega_i^2,$$

$$\text{where } d_i^2 = \frac{b_i^2\,\tilde{m}_i}{\sum_{j=1}^n b_j^2\,\tilde{m}_j}.$$

Note that $\sum_{i=1}^n d_i^2 = 1$ and that ω^2 given by the above result take values in the range $[\omega_1^2, \omega_n^2]$ since

$$\omega^2 = \sum_{i=1}^n d_i^2\,\omega_i^2 \le \sum_{i=1}^n d_i^2\,\omega_n^2 = \omega_n^2 \quad \text{and} \quad \omega^2 = \sum_{i=1}^n d_i^2\,\omega_i^2 \ge \sum_{i=1}^n d_i^2\,\omega_1^2 = \omega_1^2,$$

which proves Eq. 4.12.

Consider now Eq. 4.13. The guess $\mathbf{u} = \mathbf{\Phi}_i + \varepsilon\,\mathbf{v}$ is not far from the target modal shape since ε is small by assumption. The output of Eq. 4.11 is

$$\omega(\varepsilon)^2 = \frac{\mathbf{\Phi}_i^T\,\mathbf{k}\,\mathbf{\Phi}_i + f(\varepsilon)}{\mathbf{\Phi}_i^T\,\mathbf{m}\,\mathbf{\Phi}_i + g(\varepsilon)} = \frac{\tilde{k}_i + f(\varepsilon)}{\tilde{m}_i + g(\varepsilon)},$$

where $f(\varepsilon) = \varepsilon\left(\mathbf{v}^T\,\mathbf{k}\,\mathbf{\Phi}_i + \mathbf{\Phi}_i^T\,\mathbf{k}\,\mathbf{v}\right) + \varepsilon^2\,\mathbf{v}^T\,\mathbf{k}\,\mathbf{v}$ and $g(\varepsilon) = \varepsilon\left(\mathbf{v}^T\,\mathbf{m}\,\mathbf{\Phi}_i + \mathbf{\Phi}_i^T\,\mathbf{m}\,\mathbf{v}\right) + \varepsilon^2\,\mathbf{v}^T\,\mathbf{m}\,\mathbf{v}$. The first order Taylor's expansion of $\omega(\varepsilon)^2$ about $\varepsilon = 0$ gives

$$\omega(\varepsilon)^2 \simeq \frac{\tilde{k}_i}{\tilde{m}_i} + \frac{f'(\varepsilon)\,(\tilde{m}_i + g(\varepsilon)) - (\tilde{k}_i + f(\varepsilon))\,g'(\varepsilon)}{(\tilde{m}_i + g(\varepsilon))^2}\Big|_{\varepsilon=0}$$

$$\varepsilon = \frac{\tilde{k}_i}{\tilde{m}_i} + O(\varepsilon) = \omega_i^2 + O(\varepsilon),$$

which shows that errors of order ε in $\mathbf{\Phi}_i$ are mapped into errors of the same order of magnitude in ω_i^2.

Example 4.1 Consider the 2-DOF system in Fig. 4.2 with mass and stiffness matrices

$$\mathbf{m} = \frac{m\,h}{4}\begin{bmatrix} 2 & 0 \\ 0 & 1 \end{bmatrix} \quad \text{and} \quad \mathbf{k} = \frac{48\,E\,I}{7\,h^3}\begin{bmatrix} 16 & -5 \\ -5 & 2 \end{bmatrix},$$

where m denotes mass per unit length. The modal frequencies and shapes of the 2-DOF representation in the figure are

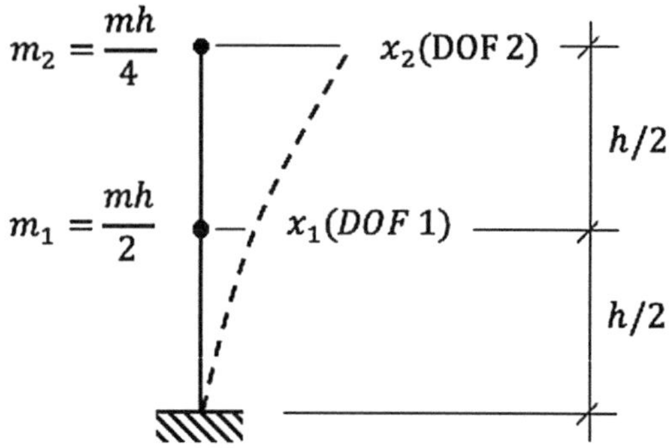

Fig. 4.2 Discrete approximation of a continuous system

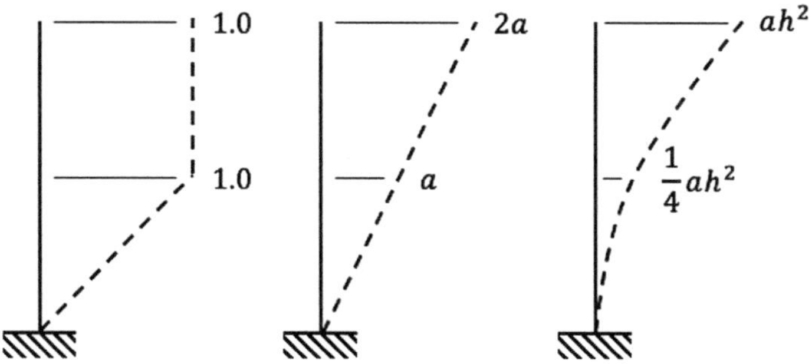

Fig. 4.3 Trials 1, 2, and 3

$$\omega_1 = 3.156\sqrt{\frac{E\,I}{m\,h^4}}, \quad \omega_2 = 16.258\sqrt{\frac{E\,I}{m\,h^4}}, \quad \mathbf{\Phi}_1 = \begin{bmatrix} 0.3111 \\ 0.9504 \end{bmatrix} \quad \text{and}$$

$$\mathbf{\Phi}_2 = \begin{bmatrix} 0.8366 \\ -0.5477 \end{bmatrix}.$$

We approximate ω_1 from Eq. 4.13 with $\mathbf{u} = [1\ 1]^T$ (Trial 1), $\mathbf{u} = [a\ 2\,a]^T, a > 0$, (Trial 2), and $\mathbf{u} = [a\,h^2/4\ a\,h^2]^T, a > 0$, (Trial 3), see Fig. 4.3. The approximations of ω_1 for these trials are

$$\omega_1 \simeq 8.5524\sqrt{\frac{E\,I}{m\,h^4}} \quad \text{(error: } +171\%\text{)},$$

$$\omega_1 \simeq 4.2763\sqrt{\frac{E\,I}{m\,h^4}}, \quad \text{(error: } +35\%\text{)},$$

$$\omega_1 \simeq 3.4913\sqrt{\frac{E\,I}{m\,h^4}} \quad \text{(error: } +10.6\%\text{)}.$$

Note that the approximations overestimate ω_1 and improve as the trial deformation $\mathbf{u}$ is closer to the first mode of vibration $\mathbf{\Phi}_1$. That our approximations overestimate ω_1 is expected since, for $\mathbf{u}$ to be the actual deformation, we need to add geometrical constraints that would make the structure stiffer.

4.3.4 Proportional Damping

Generally, the (n, n)-matrix $\mathbf{\Psi}^T\, \mathbf{c}\, \mathbf{\Psi}$ is not diagonal, where $\mathbf{c}$ is the (n, n)-damping matrix in Eq. 4.2. There is no reason to expect that the matrix $\mathbf{\Psi}^T\, \mathbf{c}\, \mathbf{\Psi}$ is diagonal since the modal shapes do not depend on $\mathbf{c}$. If $\mathbf{\Psi}^T\, \mathbf{c}\, \mathbf{\Psi}$ is diagonal, the damping matrix $\mathbf{c}$ is said to be ***proportional***. Otherwise, $\mathbf{c}$ is said to be a ***non-proportional*** damping matrix.

To simplify the analysis, it is common to assume that the damping matrix $\mathbf{c}$ is proportional. A broad range of damping matrix models have been proposed, but perhaps the most popular is the Rayleigh model that has been introduced many years ago [3]. The model assumes that the damping matrix is a weighted sum of the mass and stiffness matrices, i.e.,

$$\mathbf{c} = \alpha\, \mathbf{m} + \beta\, \mathbf{k}, \tag{4.14}$$

where $\alpha, \beta \geq 0$ are coefficients that have to be selected. The matrix $\mathbf{\Psi}^T\, \mathbf{c}\, \mathbf{\Psi}$ with $\mathbf{c}$ in Eq. 4.14 is diagonal since

$$\begin{aligned} \mathbf{\Psi}^T\, \mathbf{c}\, \mathbf{\Psi} &= \mathbf{\Psi}^T \left(\alpha\, \mathbf{m} + \beta\, \mathbf{k}\right) \mathbf{\Psi} = \alpha\, \mathbf{\Psi}^T\, \mathbf{m}\, \mathbf{\Psi} + \beta\, \mathbf{\Psi}^T\, \mathbf{k}\, \mathbf{\Psi} \\ &= \alpha\, \text{diag}\{\tilde{m}_i\} + \beta\, \text{diag}\{\tilde{k}_i\} = \text{diag}\{\tilde{c}_i\}, \end{aligned} \tag{4.15}$$

where

$$\tilde{c}_i = \alpha\, \tilde{m}_i + \beta\, \tilde{k}_i. \tag{4.16}$$

With the notation $\{2\, \zeta_i\, \omega_i = \tilde{c}_i / \tilde{m}_i\}$, $i = 1, \ldots, n$, which is similar to that used for single degree of freedom systems, we have

$$2\, \zeta_i\, \omega_i = \tilde{c}_i / \tilde{m}_i = \alpha + \beta\, \omega_i^2 \Longrightarrow \zeta_i = \frac{\alpha}{2\, \omega_i} + \frac{\beta\, \omega_i}{2}, \quad i = 1, \ldots, n. \tag{4.17}$$

According to this model, the ***modal damping ratios*** $\{\zeta_i\}$ depend on two parameters, the coefficients α and β in the definition of $\mathbf{c}$. This means that we can specify modal damping ratios for only two modes. The other modal damping ratios result since the coefficients α and β are determined by the selection of two modal damping ratios.

Example 4.2 The mass and stiffness matrices of a 3-DOF system are

$$\mathbf{m} = (1/386)\begin{bmatrix} 400 & 0 & 0 \\ 0 & 400 & 0 \\ 0 & 0 & 200 \end{bmatrix} \quad \text{and } \mathbf{k} = 610\begin{bmatrix} 2 & -1 & 0 \\ -1 & 2 & -1 \\ 0 & -1 & 1 \end{bmatrix}$$

in kips and inch units. The modal frequencies are $\omega_1 = 12.57$, $\omega_2 = 34.33$, and $\omega_3 = 46.89$ rad/s. The corresponding modal shapes are

$$\boldsymbol{\Phi}_1 = \begin{bmatrix} 0.401 \\ 0.695 \\ 0.803 \end{bmatrix}, \quad \boldsymbol{\Phi}_2 = \begin{bmatrix} 0.803 \\ 0.0 \\ -0.803 \end{bmatrix} \quad \text{and } \boldsymbol{\Phi}_3 = \begin{bmatrix} 0.401 \\ -0.695 \\ 0.803 \end{bmatrix}.$$

Suppose we set $\zeta_1 = \zeta_2 = 0.05$, i.e., we specify the damping ratios for the first two modes. The solution of the system of equations $2\,\zeta_i\,\omega_i = \alpha + \beta\,\omega_i^2$, $i = 1, 2$, in Eq. 4.17 is $\alpha = 0.9201$ and $\beta = 0.0021$. The modal damping ζ_3 cannot be selected. Its value is completely determined by Eq. 4.17 with α and β corresponding to our selection of the damping ratios ζ_1 and ζ_2. It is $\zeta_3 = \alpha/(2\,\omega_3) + \beta\,\omega_3/2 = 0.0689$.

4.3.5 Displacement Vector Representations

The following two facts are used to construct the solution of $\mathbf{x}(t)$ of Eq. 4.2 with proportional damping, i.e., the displacement of MDOF systems with proportional damping.

1. The displacement $\mathbf{x}(t)$ is an n-dimensional vector at any time t, i.e., an element of the Euclidian space $\mathbb{R}^n$. This means that $\mathbf{x}(t)$ can be represented by its projections on any system of coordinates of this space, e.g.,

$$\mathbf{x}(t) = \sum_{i=1}^{n} x_i(t)\,\mathbf{i}_i, \quad t \geq 0, \tag{4.18}$$

   where $\mathbf{i}_i$ are the unit vectors of $\mathbb{R}^n$, e.g., $\mathbf{i}_1 = (1, 0, 0, \cdots, 0)$ and $\mathbf{i}_2 = (0, 1, 0 \cdots, 0)$, and $\{x_i(t)\}$ denote the projections of $\mathbf{x}(t)$ on the unit vectors $\{\mathbf{i}_i\}$. This is a valid representation of the displacement $\mathbf{x}(t)$ but not useful for our objective, which is to break the system of coupled equations in Eq. 4.2 into a set of uncoupled equations for the components of the displacement vector $\mathbf{x}(t)$.
2. The modal shapes $\{\boldsymbol{\Phi}_i\}$ provide a basis for $\mathbb{R}^n$ if the modal frequencies $\{\omega_i\}$ are distinct (see Property 3, Sect. 3.2.1). If the modal frequencies are not distinct, we need to use generalized modal shapes (eigenvectors) to construct a basis for $\mathbb{R}^n$ (see Property 5, Sect. 3.2.1, and Appendix C). For simplicity, suppose the modal frequencies are distinct. Then, the modal shapes $\{\boldsymbol{\Phi}_i\}$ define a basis in $\mathbb{R}^n$ so that $\mathbf{x}(t)$ can be represented at any time t by

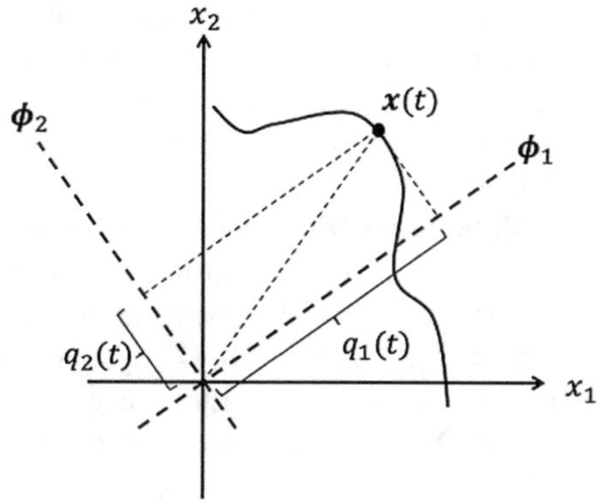

Fig. 4.4 Representation of the displacement vector $\mathbf{x}(t)$ in the coordinates defined by modal shapes

$$\mathbf{x}(t) = \sum_{i=1}^{n} \mathbf{\Phi}_i \, q_i(t) = \mathbf{\Psi} \, \mathbf{q}(t), \tag{4.19}$$

where $q_i(t)$ is the projection of $\mathbf{x}(t)$ on the coordinate $\mathbf{\Phi}_i$ of $\mathbb{R}^n$, i.e., the ith modal shape, and $\mathbf{q}(t) = [q_1(t)\, q_2(t) \ldots q_n(t)]^T$ is an n-dimensional column vector at each time t, see illustration in Fig. 4.4 for $n = 2$. Note that, if the mass matrix $\mathbf{m}$ is proportional to the identity matrix, the orthogonality condition $\mathbf{\Phi}_i^T \, \mathbf{m} \, \mathbf{\Phi}_j = 0$, $i \neq j$, of Eq. 4.9 becomes $\mathbf{\Phi}_i^T \, \mathbf{\Phi}_j = 0$, $i \neq j$, so that the modal shapes $\{\mathbf{\Phi}_i\}$ are orthogonal in the sense of the classical definition of orthogonality in $\mathbb{R}^n$.

We show that the representation of $\mathbf{x}(t)$ given by Eq. 4.19 breaks the system of coupled equations of Eq. 4.2 into a set of n independent equations for its projections $\{q_i(t)\}$ on the modal shapes and that the functional form of these equations is that of SDOF systems. The method works for MDOF systems with proportional damping. A conceptually similar method is discussed in a subsequent section for MDOF systems with non-proportional damping.

Note that the representation of the system displacement $\mathbf{x}(t)$ of Eq. 4.19, which we adopt for solving Eq. 4.2, constitutes a superposition of ***modal displacements*** $\{\mathbf{\Phi}_i \, q_i(t)\}$, i.e., time-invariant modal shapes $\{\mathbf{\Phi}_i\}$ scaled by time-dependent projections $\{q_i(t)\}$ of $\mathbf{x}(t)$ on modal shapes.

4.4 Time Domain Analysis: Proportional Damping

Consider a MDOF system with n degrees of freedom with displacement vector $\mathbf{x}(t)$ defined by Eq. 4.2. We develop methods for calculating $\mathbf{x}(t)$ for systems with proportional damping, i.e., systems with damping matrices $\mathbf{c}$ such that

$$\mathbf{\Psi}^T \, \mathbf{c} \, \mathbf{\Psi} = \text{diag}\{\tilde{c}_i\} \tag{4.20}$$

is a diagonal matrix with non-zero entries $\{\tilde{c}_i\}$.

The presentation is organized as follows. Methods for calculating the forced vibrations of damped MDOF systems are discussed in Sect. 4.4.1. These methods are applied in Sect. 4.4.2 to find the response of MDOF to seismic events. The remaining subsections are special cases of that in Sect. 4.4.1. They first suggest how to specialize the results of Sect. 4.4.1 and then, for completeness, construct the solutions for the special cases under consideration by direct arguments. The free vibration of undamped MDOF systems is examined in Sect. 4.4.3. Torsional vibration in Sect. 4.4.4 provides an illustration for these systems. The forced vibration of undamped MDOF systems is considered in Sect. 4.4.5. Experimental estimation of modal frequencies in Sect. 4.4.6 provides an application for this case. The free vibration of damped systems is examined in Sect. 4.4.7.

4.4.1 Damped Systems: Forced Vibration

We have seen that the displacement vector $\mathbf{x}(t)$ defined by Eq. 4.2 with initial displacement $\mathbf{x}_0$ and velocity $\dot{\mathbf{x}}_0$ admits the representation in Eq. 4.19 since it is the element of $\mathbb{R}^n$ at any time $t \geq 0$ and the eigenvectors $\{\boldsymbol{\Phi}_i\}$ span this space. We have seen that the eigenvectors $\{\boldsymbol{\Phi}_i\}$ of this representation can be obtained from the mass and stiffness matrices of the dynamical system under consideration. To find the projections $\{q_i(t)\}$ of $\mathbf{x}(t)$ on the eigenvectors $\{\boldsymbol{\Phi}_i\}$, referred to as ***modal coordinates***, we require that $\mathbf{x}(t)$ given by Eq. 4.19 satisfies the equation of motion Eq. 4.2, i.e.,

$$\mathbf{m}\,\boldsymbol{\Psi}\,\ddot{\mathbf{q}}(t) + \mathbf{c}\,\boldsymbol{\Psi}\,\dot{\mathbf{q}}(t) + \mathbf{k}\,\boldsymbol{\Psi}\,\mathbf{q}(t) = \mathbf{f}(t). \tag{4.21}$$

To break this system of coupled equations into a set of independent equations for the modal coordinates $\{q_i(t)\}$, we use the orthogonality of modal shapes. The left multiplication of the above equation by $\boldsymbol{\Psi}^T$ gives

$$\boldsymbol{\Psi}^T\,\mathbf{m}\,\boldsymbol{\Psi}\,\ddot{\mathbf{q}}(t) + \boldsymbol{\Psi}^T\,\mathbf{c}\,\boldsymbol{\Psi}\,\dot{\mathbf{q}}(t) + \boldsymbol{\Psi}^T\,\mathbf{k}\,\boldsymbol{\Psi}\,\mathbf{q}(t) = \boldsymbol{\Psi}^T\,\mathbf{f}(t), \tag{4.22}$$

or

$$\mathrm{diag}\{\tilde{m}_i\}\,\ddot{\mathbf{q}}(t) + \boldsymbol{\Psi}^T\,\mathbf{c}\,\boldsymbol{\Psi}\,\dot{\mathbf{q}}(t) + \mathrm{diag}\{\tilde{k}_i\}\,\mathbf{q}(t) = \boldsymbol{\Psi}^T\,\mathbf{f}(t) \tag{4.23}$$

by the orthogonality of the modal shapes (see Eq. 4.10), where $\mathrm{diag}\{\tilde{m}_i\}$ and $\mathrm{diag}\{\tilde{k}_i\}$ are (n, n)-diagonal matrices whose non-zero entries are the modal masses $\{m_i\}$ and modal stiffnesses $\{k_i\}$.

Generally, the matrix $\boldsymbol{\Psi}^T\,\mathbf{c}\,\boldsymbol{\Psi}$ is not diagonal so that it is not possible to decouple the system of equations of motion. However, it is diagonal under the assumption of proportional damping in Eq. 4.20. Under this assumption, Eq. 4.23 becomes

$$\mathrm{diag}\{\tilde{m}_i\}\,\ddot{\mathbf{q}}(t) + \mathrm{diag}\{\tilde{c}_i\}\,\dot{\mathbf{q}}(t) + \mathrm{diag}\{\tilde{k}_i\}\,\mathbf{q}(t) = \boldsymbol{\Psi}^T\,\mathbf{f}(t), \tag{4.24}$$

i.e., the set of n independent equations

$$\tilde{m}_i\, \ddot{q}_i(t) + \tilde{c}_i\, \ddot{q}_i(t) + \tilde{k}_i\, q_i(t) = \left(\boldsymbol{\Psi}^T\, \mathbf{f}(t)\right)_i, \quad i = 1, \ldots, n, \tag{4.25}$$

for the modal coordinates $\{q_i(t)\}$, which can be given in the form

$$\ddot{q}_i(t) + 2\,\zeta_i\,\omega_i\,\ddot{q}_i(t) + \omega_i^2\, q_i(t) = f_i(t)/\tilde{m}_i, \quad i = 1, \ldots, n, \tag{4.26}$$

where $f_i(t) = \left(\boldsymbol{\Psi}^T\, \mathbf{f}(t)\right)_i$ denotes the ith component of the n-dimensional vector $\boldsymbol{\Psi}^T\, \mathbf{f}(t)$, $\tilde{c}_i/\tilde{m}_i = 2\,\zeta_i\,\omega_i$, ζ_i is called ***modal damping ratio***, and $\tilde{k}_i/\tilde{m}_i = \omega_i^2$ (see Eq. 4.9).

The latter equations show that the modal coordinates $\{q_i(t)\}$ satisfy differential equations that are analogous to those of damped SDOF systems with masses $\{\tilde{m}_i\}$, natural frequencies $\{\omega_i\}$, and damping ratios $\{\zeta_i\}$ under forcing functions $\{f_i(t)\}$. For modal damping ratios $\zeta_i < 1$, the modal coordinates can be calculated from

$$q_i(t) = e^{-\zeta_i\,\omega_i\,t}\left(q_{0,i}\,\cos(\omega_{d,i}\,t) + \frac{\dot{q}_{0,i} + \zeta_i\,\omega_i\,q_{0,i}}{\omega_{d,i}}\,\sin(\omega_{d,i}\,t)\right)$$
$$+ \int_0^t h_i(t-u)\, f_i(u)\, du, \tag{4.27}$$

where $\omega_{d,i} = \omega_i\,\sqrt{1-\zeta_i^2}$ and

$$h_i(s) = \frac{1}{\tilde{m}_1\,\omega_{d,i}}\, e^{-\zeta_i\,\omega_i\,s}\,\sin\left(\omega_{d,i}\,s\right), \quad i = 1, \ldots, n, \tag{4.28}$$

denote the unit impulse response function of the ith mode of vibration (see Eqs. 2.13, 2.21 and 2.30). The displacement vector has the form

$$\mathbf{x}(t) = \sum_{i=1}^{n} \boldsymbol{\Phi}_i \left[e^{-\zeta_i\,\omega_i\,t}\left(q_{0,i}\,\cos(\omega_{d,i}\,t) + \frac{\dot{q}_{0,i} + \zeta_i\,\omega_i\,q_{0,i}}{\omega_{d,i}}\,\sin(\omega_{d,i}\,t)\right)\right.$$
$$\left. + \int_0^t h_i(t-u)\, f_i(u)\, du\right]. \tag{4.29}$$

Since $0 < \zeta_i < 1$, $i = 1, \ldots, n$, the free vibration component of the displacement vanishes as time increases indefinitely, so that the system displacement becomes

$$\mathbf{x}_{SS}(t) \simeq \sum_{i=1}^{n} \boldsymbol{\Phi}_i \int_0^t h_i(t-u)\, f_i(u)\, du \tag{4.30}$$

for large times. It is referred to as the ***steady-state*** displacement vector.

4.4.1.1 Initial Conditions for $\mathbf{q}(t)$

The initial conditions for the modal coordinates result from the initial conditions $(\mathbf{x}_0, \dot{\mathbf{x}}_0)$ in the physical space and the representation of Eq. 4.19 at the initial time $t = 0$. Two methods are commonly used to find the initial conditions $\mathbf{q}_0$ and $\dot{\mathbf{q}}_0$ for the solution $\mathbf{q}(t)$ of Eq. 4.25 and Eq. 4.26 from the specified initial displacement $\mathbf{x}_0$ and the initial velocity $\dot{\mathbf{x}}_0$ of $\mathbf{x}(t)$.

- *Method 1:* It uses the relationship between $\mathbf{q}(t)$ and $\mathbf{x}(t)$ in Eq. 4.19, which holds at all times, so that we have $\mathbf{x}(0) = \mathbf{x}_0 = \boldsymbol{\Psi}\,\mathbf{q}(0)$ and $\dot{\mathbf{x}}(0) = \dot{\mathbf{x}}_0 = \boldsymbol{\Psi}\,\dot{\mathbf{q}}(0)$. The inversions of these relationships,

$$\mathbf{q}_0 = \boldsymbol{\Psi}^{-1}\,\mathbf{x}_0 \quad \text{and} \quad \dot{\mathbf{q}}_0 = \boldsymbol{\Psi}^{-1}\,\dot{\mathbf{x}}_0, \tag{4.31}$$

  give the required initial conditions.
- *Method 2:* It uses the orthogonality of modal shapes. For example, the left multiplication of $\mathbf{x}_0 = \boldsymbol{\Psi}\,\mathbf{q}_0$ by $\boldsymbol{\Psi}^T\,\mathbf{m}$ gives $\boldsymbol{\Psi}^T\,\mathbf{m}\,\mathbf{x}_0 = \boldsymbol{\Psi}^T\,\mathbf{m}\,\boldsymbol{\Psi}\,\mathbf{q}_0 = \text{diag}\{\tilde{m}_i\}\,\mathbf{q}_0 = [\tilde{m}_1\,q_{1,0}\cdots\tilde{m}_i\,q_{i,0}\cdots\tilde{m}_n\,q_{n,0}]^T$, so that $\tilde{m}_i\,q_{i,0}$ is equal to the ith component of the n-dimensional vector $\boldsymbol{\Psi}^T\,\mathbf{m}\,\mathbf{x}_0$. Similarly, we have that $\tilde{m}_i\,\dot{q}_{i,0}$ is equal to the ith component of the n-dimensional vector $\boldsymbol{\Psi}^T\,\mathbf{m}\,\dot{\mathbf{x}}_0$. In summary, we have

$$q_{i,0} = \frac{\left(\boldsymbol{\Psi}^T\,\mathbf{m}\,\mathbf{x}_0\right)_i}{\tilde{m}_i} \quad \text{and} \quad \dot{q}_{i,0} = \frac{\left(\boldsymbol{\Psi}^T\,\mathbf{m}\,\dot{\mathbf{x}}_0\right)_i}{\tilde{m}_i}, \quad i = 1, \ldots, n. \tag{4.32}$$

4.4.1.2 System Displacement $\mathbf{x}(t)$

To conclude, the construction of the solution $\mathbf{x}(t)$ involves the following three steps.

- *Step 1. Modal analysis:* Find the modal shapes and frequencies of the system from, e.g., the output of the **eig** MATLAB function with argument the (n, n)-matrix $\mathbf{m}^{-1}\,\mathbf{k}$ (see Eq. 4.8).
- *Step 2. Modal coordinates:* Use the modal shapes and frequencies to construct the differential equations of motion for the modal coordinates $\{q_i(t)\}$ (see Eq. 4.26), find the initial conditions for these equations, and solve these equations by the methods developed for SDOF systems in Sect. 2.4 (time domain analysis) and Sect. 2.5 (frequency domain analysis).
- *Step 3. System displacement:* Use the representation of the system displacement in Eq. 4.19 to construct the displacement vector $\mathbf{x}(t)$.

Example 4.3 Consider a 2-DOF system with mass, stiffness, and damping matrices

$$\mathbf{m} = \begin{bmatrix} 2 & 0 \\ 0 & 1 \end{bmatrix}, \quad \mathbf{k} = \begin{bmatrix} 16 & -5 \\ -5 & 2 \end{bmatrix},$$

and $\mathbf{c} = 0.0505\,\mathbf{m} + 0.0270\,\mathbf{k}$, which is at rest at the initial time, i.e., $\mathbf{x}_0 = \mathbf{0}$ and $\dot{\mathbf{x}}_0 = \mathbf{0}$. The modal frequencies and modal shapes of the system are $\omega_1 = 0.6027$ and $\omega_2 = 3.1043$ and

$$\boldsymbol{\Psi} = \begin{bmatrix} 0.3111 & 0.8366 \\ 0.9504 & -0.5478 \end{bmatrix}.$$

The system is subjected to the harmonic force $f(t) = \sin(\nu\, t)$ applied at the second degree of freedom so that

$$\mathbf{f}(t) = \begin{bmatrix} 0 \\ 1 \end{bmatrix} \sin(\nu\, t).$$

The modal coordinates are solutions of Eq. 4.26 with zero initial conditions since $\mathbf{x}_0 = \mathbf{0}$ and $\dot{\mathbf{x}}_0 = \mathbf{0}$ (see Eqs. 4.31 or 4.32). The forcing functions $f_i(t) = \left(\boldsymbol{\Psi}^T\,\mathbf{f}(t)\right)_i$ are $f_i(t) = \Phi_{i,2}\,\sin(\nu\, t)$, $i = 1, 2$, where $\Phi_{i,r}$ denotes the rth component of $\boldsymbol{\Phi}_i$. The modal coordinate can be obtained from Eq. 4.27 or results for SDOF systems given in Sect. 2.4.5.

The left panel of Fig. 4.5 shows with solid and dashed lines the first and second modal responses $\boldsymbol{\Phi}_1\, q_1(t)$ and $\boldsymbol{\Phi}_2\, q_2(t)$ for $\nu = 2$. The right panel shows the system displacements $x_1(t)$ and $x_2(t)$. Similar plots are in Fig. 4.6 for $\nu = 3$. The system displacements $\{x_i(t)\}$ for $\nu = 2$ are dominated by the first mode response as it can be seen by comparing modal and system responses in the two panels of the figure. This observation is also supported by the fact that the displacements $x_1(t)$ and $x_2(t)$ have the same sign at all times. For $\nu = 3$, the second mode response becomes dominant in time since the forcing frequency ν and modal frequency ω_2 are nearly equal. Note also that the signs of the displacements $x_1(t)$ and $x_2(t)$ differ at times, which indicates that the second mode contributes significantly to the overall system displacement.

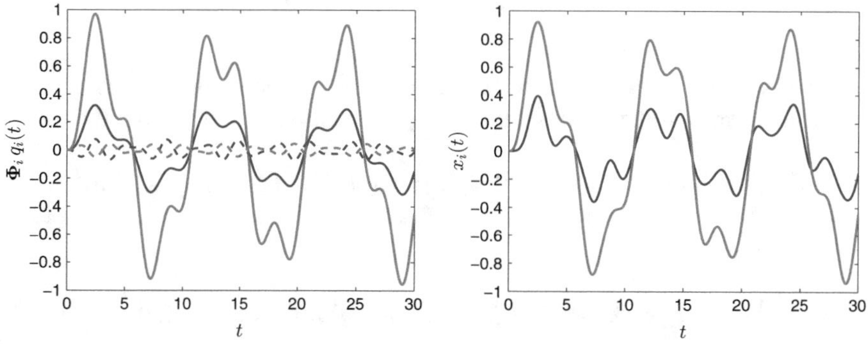

Fig. 4.5 Modal responses $\boldsymbol{\Phi}_1\, q_1(t)$ and $\boldsymbol{\Phi}_2\, q_2(t)$ (solid and dashed lines) for $\nu = 2$ and the corresponding system displacements $\{x_i(t)\}$ (left and right panels)

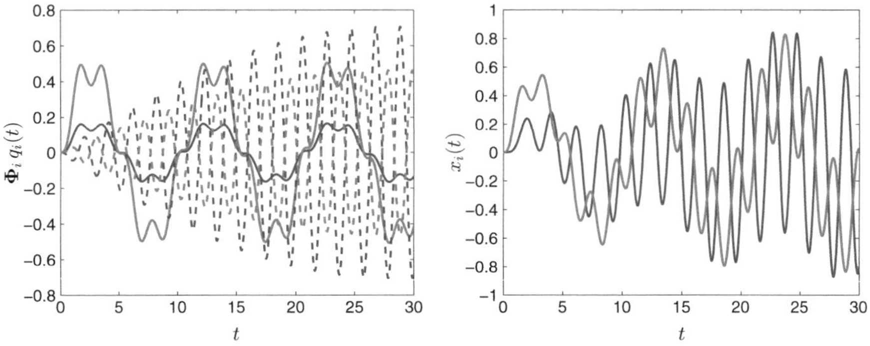

Fig. 4.6 Modal responses $\mathbf{\Phi}_1\, q_1(t)$ and $\mathbf{\Phi}_2\, q_2(t)$ (solid and dashed lines) for $\nu = 3$ and the corresponding system displacements $\{x_i(t)\}$ (left and right panels)

4.4.1.3 Solution Uniqueness

We have seen that the modal shapes can be determined up to a multiplicative constant so that it makes sense to question the validity of the modal analysis method discussed in this section. It is necessary to show that the displacement vector $\mathbf{x}(t)$ is unique in the sense that it does not depend on the particular scaling of the modal shapes used in analysis. The following arguments show that we obtained the same displacement vector $\mathbf{x}(t)$ irrespective of modal shape scaling.

Consider the representation of $\mathbf{x}(t)$ in Eq. 4.19, which implies that the modal coordinates are the solutions of Eq. 4.26 with initial conditions given by, e.g., Eq. 4.31. Suppose that a different scaling is used for the modal shapes, i.e., the original modes $\{\mathbf{\Phi}_i\}$ are replaced with $\{e_i\, \mathbf{\Phi}_i\}$, where $\{e_i\}$ are arbitrary scaling constants. Accordingly, the representation of the solution $\mathbf{x}(t)$ in Eq. 4.19, denoted temporarily by $\mathbf{y}(t)$, becomes

$$\mathbf{y}(t) = \sum_{i=1}^{n} \left(e_i\, \mathbf{\Phi}_i\right) p_i(t) = \mathbf{\Psi}\, \mathbf{e}\, \mathbf{p}(t), \tag{4.33}$$

where $\mathbf{e}$ is an (n, n)-diagonal matrix with non-zero entries $\{e_i\}$ and $\mathbf{p}(t)$ is an n-dimensional vector. The modal coordinates $\mathbf{p}(t)$ are the solutions of (see Eq. 4.22)

$$\left(\mathbf{\Psi}\, \mathbf{e}\right)^T \mathbf{m} \left(\mathbf{\Psi}\, \mathbf{e}\right) \ddot{\mathbf{p}}(t) + \left(\mathbf{\Psi}\, \mathbf{e}\right)^T \mathbf{c} \left(\mathbf{\Psi}\, \mathbf{e}\right) \dot{\mathbf{p}}(t) + \left(\mathbf{\Psi}\, \mathbf{e}\right)^T \mathbf{k} \left(\mathbf{\Psi}\, \mathbf{e}\right) \mathbf{p}(t) = \left(\mathbf{\Psi}\, \mathbf{e}\right)^T \mathbf{f}(t), \tag{4.34}$$

which gives (see Eq. 4.24)

$$\operatorname{diag}\{e_i^2\, \tilde{m}_i\}\, \ddot{\mathbf{p}}(t) + \operatorname{diag}\{e_i^2\, \tilde{c}_i\}\, \dot{\mathbf{p}}(t) + \operatorname{diag}\{e_i^2\, \tilde{k}_i\}\, \mathbf{p}(t) = \mathbf{e}^T\, \mathbf{\Psi}^T\, \mathbf{f}(t), \tag{4.35}$$

so that

$$e_i^2\,\tilde{m}_i\,\ddot{p}_i(t) + e_i^2\,\tilde{c}_i\,\dot{p}_i(t) + e_i^2\,\tilde{k}_i\,p_i(t) = e_i\left(\boldsymbol{\Psi}^T\,\mathbf{f}(t)\right)_i, \quad \text{or}$$
$$\tilde{m}_i\,\ddot{p}_i(t) + \tilde{c}_i\,\dot{p}_i(t) + \tilde{k}_i\,p_i(t) = \left(\boldsymbol{\Psi}^T\,\mathbf{f}(t)\right)_i/e_i, \quad i = 1,\ldots,n, \tag{4.36}$$

since, e.g.,

$$\left(\boldsymbol{\Psi}\,\mathbf{e}\right)^T\,\mathbf{m}\left(\boldsymbol{\Psi}\,\mathbf{e}\right) = \left(\mathbf{e}^T\,\boldsymbol{\Psi}^T\,\mathbf{m}\,\boldsymbol{\Psi}\,\mathbf{e}\right) = \mathbf{e}^T\,\mathrm{diag}\{\tilde{m}_i\}\,\mathbf{e} = \mathrm{diag}\{e_i^2\,\tilde{m}_i\} \quad \text{and}$$
$$\left(\mathbf{e}^T\,\boldsymbol{\Psi}^T\,\mathbf{f}(t)\right)_i = e_i\left(\boldsymbol{\Psi}^T\,\mathbf{f}(t)\right)_i, \quad i = 1,\ldots,n. \tag{4.37}$$

The initial conditions for Eq. 4.36 are (see, e.g., Eq. 4.31)

$$\mathbf{p}_0 = \left(\boldsymbol{\Psi}\,\mathbf{e}\right)^{-1}\mathbf{x}_0 = \mathbf{e}^{-1}\,\boldsymbol{\Psi}^{-1}\,\mathbf{x}_0 = \mathrm{diag}\{1/e_i\}\,\boldsymbol{\Psi}^{-1}\,\mathbf{x}_0 = \mathrm{diag}\{1/e_i\}\,\mathbf{q}_0 \quad \text{and}$$
$$\dot{\mathbf{p}}_0 = \left(\boldsymbol{\Psi}\,\mathbf{e}\right)^{-1}\dot{\mathbf{x}}_0 = \mathbf{e}^{-1}\,\boldsymbol{\Psi}^{-1}\,\dot{\mathbf{x}}_0 = \mathrm{diag}\{1/e_i\}\,\boldsymbol{\Psi}^{-1}\,\dot{\mathbf{x}}_0 = \mathrm{diag}\{1/e_i\}\,\dot{\mathbf{q}}_0. \tag{4.38}$$

The comparison of the latter equation in Eq. 4.36 satisfying the initial conditions of Eq. 4.38 with the equation of motion for the original modal coordinates $\{q_i(t)\}$ in Eq. 4.26 satisfying the initial conditions of Eq. 4.31 shows that $p_i(t) = q_i(t)/e_i$, $i = 1,\ldots,n$, so that $\mathbf{y}(t)$ in Eq. 4.33 is

$$\mathbf{y}(t) = \sum_{i=1}^{n}\left(e_i\,\boldsymbol{\Phi}_i\right)p_i(t) = \sum_{i=1}^{n}\left(e_i\,\boldsymbol{\Phi}_i\right)q_i(t)/e_i = \boldsymbol{\Psi}\,\mathbf{q}(t) = \mathbf{x}(t), \tag{4.39}$$

which shows that the system displacement does not depend on the scaling used for modal shapes.

4.4.2 Damped Systems: Free Vibration

The equation of motion is given by Eq. 4.2 with $\mathbf{c} \neq \mathbf{0}$, and $\mathbf{f}(t) = \mathbf{0}$ becomes

$$\mathbf{m}\,\ddot{\mathbf{x}} + \mathbf{c}\,\dot{\mathbf{x}} + \mathbf{k}\,\mathbf{x} = \mathbf{0}, \tag{4.40}$$

where $\mathbf{m}$, $\mathbf{c}$, and $\mathbf{k}$ denote the mass, damping, and stiffness matrices, and $\mathbf{x}$ is the displacement vector. The solution of Eq. 4.40 for some initial conditions $(\mathbf{x}_0, \dot{\mathbf{x}}_0)$ can be obtained from results of Sect. 4.4.1 by setting $\mathbf{f}(t) = \mathbf{0}$. For completeness, we construct the solution of Eq. 4.40 by direct arguments.

The equation of motion given by Eq. 4.40 with the representation of the displacement vector $\mathbf{x}(t)$ in Eq. 4.19 takes the form

$$\mathbf{m}\,\boldsymbol{\Psi}\,\ddot{\mathbf{q}}(t) + \mathbf{c}\,\boldsymbol{\Psi}\,\dot{\mathbf{q}}(t) + \mathbf{k}\,\boldsymbol{\Psi}\,\mathbf{q}(t) = \mathbf{0}, \tag{4.41}$$

which gives

$$\boldsymbol{\Psi}^T\, \mathbf{m}\, \boldsymbol{\Psi}\, \ddot{\mathbf{q}}(t) + \boldsymbol{\Psi}^T\, \mathbf{c}\, \boldsymbol{\Psi}\, \dot{\mathbf{q}}(t) + \boldsymbol{\Psi}^T\, \mathbf{k}\, \boldsymbol{\Psi}\, \mathbf{q}(t) = \mathbf{0},$$

by left multiplication with $\boldsymbol{\Psi}^T$. The orthogonality of Eq. 4.10 implies

$$\begin{aligned} &\mathrm{diag}\{\tilde{m}_i\}\, \ddot{\mathbf{q}}(t) + \mathrm{diag}\{\tilde{c}_i\}\, \dot{\mathbf{q}}(t) + \mathrm{diag}\{\tilde{k}_i\}\, \mathbf{q}(t) = \mathbf{0} \quad \text{or} \\ &\tilde{m}_i\, \ddot{q}_i(t) + \tilde{c}_i\, \ddot{q}_i(t) + \tilde{k}_i\, q_i(t) = 0, \quad i = 1, \ldots, n. \end{aligned} \tag{4.42}$$

To find the system response $\mathbf{x}(t)$, we need to calculate the free vibration solutions of n damped SDOF systems for the initial conditions $\mathbf{q}_0 = \boldsymbol{\Psi}^{-1}\, \mathbf{x}_0$, and $\dot{\mathbf{q}}_0 = \boldsymbol{\Psi}^{-1}\, \dot{\mathbf{x}}_0$. These solutions are (see Sect. 2.4.2 on SDOF systems)

$$q_i(t) = e^{-\zeta_i\, \omega_i\, t} \left[q_{i,0}\, \cos\left(\omega_{d,i}\, t\right) + \frac{\dot{q}_{i,0} + \zeta_i\, \omega_i\, q_{i,0}}{\omega_{d,i}}\, \sin\left(\omega_{d,i}\, t\right) \right], \quad i = 1, \ldots, n, \tag{4.43}$$

where $2\,\zeta_i\, \omega_i = \tilde{c}_i / \tilde{m}_i$, and $\omega_{d,i} = \omega_i \sqrt{1 - \zeta_i^2}$. The system displacement results from the representation $\mathbf{x}(t) = \sum_{i=1}^{n} \boldsymbol{\Phi}_i\, q_i(t)$ of $\mathbf{x}(t)$ and the above solutions. It has the following expression:

$$\begin{aligned} \mathbf{x}(t) = \sum_{i=1}^{n} \boldsymbol{\Phi}_i\, q_i(t) = \sum_{i=1}^{n} \boldsymbol{\Phi}_i\, e^{-\zeta_i\, \omega_i\, t} &\left[q_{i,0}\, \cos\left(\omega_{d,i}\, t\right) \right. \\ &\left. + \frac{\dot{q}_{i,0} + \zeta_i\, \omega_i\, q_{i,0}}{\omega_{d,i}}\, \sin\left(\omega_{d,i}\, t\right) \right]. \end{aligned} \tag{4.44}$$

4.4.3 Undamped Systems: Forced Vibration

The equation of motion Eq. 4.2 with $\mathbf{c} = \mathbf{0}$ and $\mathbf{f}(t) \neq \mathbf{0}$ becomes

$$\mathbf{m}\, \ddot{\mathbf{x}} + \mathbf{k}\, \mathbf{x} = \mathbf{f}(t), \tag{4.45}$$

where $\mathbf{m}$ and $\mathbf{k}$ denote the mass and stiffness matrices, $\mathbf{x}$ is the displacement vector, and $\mathbf{f}(t)$ is the n-dimensional force. The displacement vector $\mathbf{x}(t)$ results from that for the forced vibration of damped systems in Sect. 4.4.1 by setting $\zeta_i = 0$. For completeness, we construct the solution of Eq. 4.45 by direct arguments.

The equation of motion of Eq. 4.45 with the representation of the displacement vector $\mathbf{x}(t)$ in Eq. 4.19 takes the form

$$\mathbf{m}\, \boldsymbol{\Psi}\, \ddot{\mathbf{q}}(t) + \mathbf{k}\, \boldsymbol{\Psi}\, \mathbf{q}(t) = \mathbf{f}(t), \tag{4.46}$$

which gives

$$\boldsymbol{\Psi}^T \mathbf{m}\, \boldsymbol{\Psi}\, \ddot{\mathbf{q}}(t) + \boldsymbol{\Psi}^T \mathbf{k}\, \boldsymbol{\Psi}\, \mathbf{q}(t) = \boldsymbol{\Psi}^T \mathbf{f}(t),$$

by left multiplication with $\boldsymbol{\Psi}^T$. The orthogonality condition of Eq. 4.10 implies

$$\begin{aligned} &\operatorname{diag}\{\tilde{m}_i\}\, \ddot{\mathbf{q}}(t) + \operatorname{diag}\{\tilde{k}_i\}\, \mathbf{q}(t) = \boldsymbol{\Psi}^T \mathbf{f}(t) \quad \text{or} \\ &\tilde{m}_i\, \ddot{q}_i(t) + \tilde{k}_i\, q_i(t) = f_i(t), \quad i = 1, \ldots, n, \end{aligned} \tag{4.47}$$

where $f_i(t) = \left(\boldsymbol{\Psi}^T \mathbf{f}(t)\right)_i$ denotes the ith component of the n-dimensional vector $\boldsymbol{\Psi}^T \mathbf{f}(t)$. The latter equation becomes

$$\ddot{q}_i(t) + \omega_i^2\, q_i(t) = f_i(t)/\tilde{m}_i, \quad i = 1, \ldots, n, \tag{4.48}$$

by using Eq. 4.9 following division with $\tilde{m}_i$.

These equations describe the forced vibrations of n undamped SDOF systems with natural frequencies $\{\omega_i\}$ subjected to the forcing functions $\{f_i(t) = \left(\boldsymbol{\Psi}^T \mathbf{f}(t)\right)_i\}$. Their solutions have the form (see results on SDOF systems)

$$q_i(t) = q_{i,0} \cos(\omega_i\, t) + \frac{\dot{q}_{i,0}}{\omega_i} \sin(\omega_i\, t) + q_{p,i}(t), \quad i = 1, \ldots, n, \tag{4.49}$$

where $(q_{i,0}, \dot{q}_{i,0})$ are the initial conditions for these oscillators, which can be obtained from, e.g., Eq. 4.31, and $q_{p,i}(t)$ are particular solutions that depend on the types of forcing functions.

The forced vibration solution of the MDOF system has the expression

$$\mathbf{x}(t) = \sum_{i=1}^{n} \boldsymbol{\Phi}_i\, q_i(t) = \sum_{i=1}^{n} \boldsymbol{\Phi}_i \left[q_{i,0} \cos(\omega_i\, t) + \frac{\dot{q}_{i,0}}{\omega_i} \sin(\omega_i\, t) + q_{p,i}(t) \right]. \tag{4.50}$$

Example 4.4 Consider the 2-DOF structure in Fig. 4.7 in Example 4.5. The system is at rest at the initial time and is subjected to the force $\sin(\nu\, t)$, $\nu \neq \omega_i$, $i = 1, 2$, applied at the second degree of freedom, as illustrated in the figure. The motion of this system is described by Eq. 4.24, $n = 2$, with the forcing function

$$\mathbf{f}(t) = \begin{bmatrix} 0 \\ 1 \end{bmatrix} \sin(\nu\, t).$$

The calculation of the displacement vector $\mathbf{x}(t)$ involves the following three steps.

- *Step 1: Modal analysis*: The modal shapes and frequencies are calculated in Example 4.5 of the following subsection and are used here.

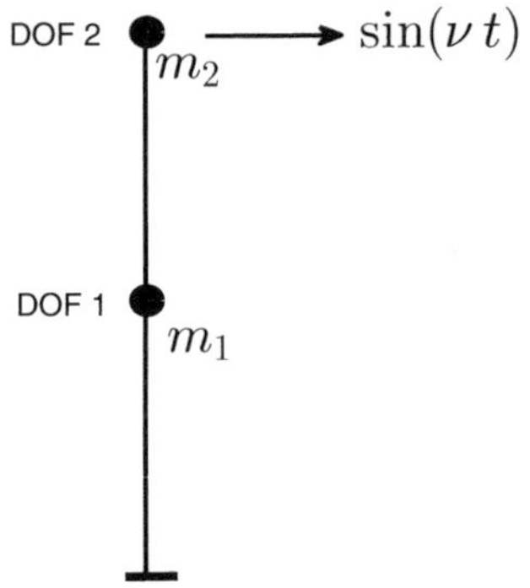

Fig. 4.7 2DOF system subjected to a sinusoidal load

– *Step 2: Modal coordinates*: From Eq. 4.47, we have

$$\tilde{m}_1\,\ddot{q}_1(t) + \tilde{k}_1\,q_1(t) = 0.9504\,\sin(\nu\,t)$$
$$\tilde{m}_2\,\ddot{q}_2(t) + \tilde{k}_2\,q_2(t) = -0.5478\,\sin(\nu\,t),$$

so that (see Sect. 2.4.5 on SDOF systems)

$$q_1(t) = \frac{0.9504/\tilde{k}_1}{1-(\nu/\omega_1)^2}\big(\sin(\nu\,t) - \frac{\nu}{\omega_1}\,\sin(\omega_1\,t)\big)$$
$$q_2(t) = \frac{-0.5478/\tilde{k}_2}{1-(\nu/\omega_2)^2}\big(\sin(\nu\,t) - \frac{\nu}{\omega_2}\,\sin(\omega_2\,t)\big).$$

– *Step 3. System solution*: $\mathbf{x}(t) = \mathbf{\Phi}_1\,q_1(t) + \mathbf{\Phi}_2\,q_2(t)$.

4.4.4 Undamped Systems: Free Vibration

The equation of motion results from Eq. 4.2 with $\mathbf{c} = \mathbf{0}$ and $\mathbf{f}(t) = \mathbf{0}$. It has the form

$$\mathbf{m}\,\ddot{\mathbf{x}} + \mathbf{k}\,\mathbf{x} = \mathbf{0}, \tag{4.51}$$

where $\mathbf{m}$ and $\mathbf{k}$ denote the mass and stiffness matrices, $\mathbf{x}$ is the displacement vector, and $(\mathbf{x}_0, \dot{\mathbf{x}}_0)$ are initial conditions. The solution $\mathbf{x}(t)$ of Eq. 4.51 for specified initial conditions $(\mathbf{x}_0, \dot{\mathbf{x}}_0)$ results from that of Sect. 4.4.1 by setting $\zeta_i = 0$ and $f_i(t) = 0$. For completeness, we also construct the solution of this equation by direct arguments.

The equation of motion of Eq. 4.51 with the representation of the displacement vector $\mathbf{x}(t)$ Eq. 4.19 takes the form

$$\mathbf{m}\,\mathbf{\Psi}\,\ddot{\mathbf{q}}(t) + \mathbf{k}\,\mathbf{\Psi}\,\mathbf{q}(t) = \mathbf{0}, \tag{4.52}$$

which gives

$$\boldsymbol{\Psi}^T \, \mathbf{m} \, \boldsymbol{\Psi} \, \ddot{\mathbf{q}}(t) + \boldsymbol{\Psi}^T \, \mathbf{k} \, \boldsymbol{\Psi} \, \mathbf{q}(t) = \mathbf{0},$$

by left multiplication with $\boldsymbol{\Psi}^T$. The orthogonality condition of Eq. 4.10 implies

$$\text{diag}\{\tilde{m}_i\} \, \ddot{\mathbf{q}}(t) + \text{diag}\{\tilde{k}_i\} \, \mathbf{q}(t) = \mathbf{0} \quad \text{or, equivalently,} \quad \tilde{m}_i \, \ddot{q}_i(t) + \tilde{k}_i \, q_i(t) = 0,$$
$$i = 1, \ldots, n$$

so that

$$\ddot{q}_i(t) + \omega_i^2 \, q_i(t) = 0, \quad i = 1, \ldots, n, \tag{4.53}$$

by using Eq. 4.9.

The differential equations of Eq. 4.53 describe the motion of n undamped single degree of freedom (SDOF) systems with natural frequencies $\{\omega_i\}$ in free vibration so that (see results for SDOF systems)

$$q_i(t) = q_{i,0} \, \cos(\omega_i \, t) + \frac{\dot{q}_{i,0}}{\omega_i} \, \sin(\omega_i \, t), \tag{4.54}$$

where $(q_{i,0}, \dot{q}_{i,0})$ are the ICs for these oscillators, which can be obtained, e.g., from Eq. 4.31 or Eq. 4.32. The free vibration solution of the MDOF system is

$$\mathbf{x}(t) = \sum_{i=1}^{n} \boldsymbol{\Phi}_i \, q_i(t) = \sum_{i=1}^{n} \boldsymbol{\Phi}_i \left(q_{i,0} \, \cos(\omega_i \, t) + \frac{\dot{q}_{i,0}}{\omega_i} \, \sin(\omega_i \, t) \right). \tag{4.55}$$

Example 4.5 Consider the cantilever in Fig. 4.8 with the masses $m_1 = 2$ and $m_2 = 1$ at the cantilever mid height and its free end. The mass and stiffness matrices of this two degree of freedom system are

$$\mathbf{m} = \begin{bmatrix} 2 & 0 \\ 0 & 1 \end{bmatrix} \quad \text{and } \mathbf{k} = \begin{bmatrix} 16 & -5 \\ -5 & 2 \end{bmatrix}.$$

Our objective is to find the system displacement $\mathbf{x}(t)$ for the initial conditions $\mathbf{x}_0 \neq \mathbf{0}$ and $\dot{\mathbf{x}}_0 = \mathbf{0}$. The following three steps deliver the displacement vector.

Step1: Modal analysis:

– *Modal frequencies:* $\det\big(-\omega^2 \mathbf{m} + \mathbf{k} \big) = 0$, i.e.,

$$\begin{bmatrix} -2\omega^2 + 16 & -5 \\ -5 & -\omega^2 + 2 \end{bmatrix} = 0 \Rightarrow \begin{array}{l} \omega_1^2 = 0.3632 \\ \omega_2^2 = 9.6968. \end{array}$$

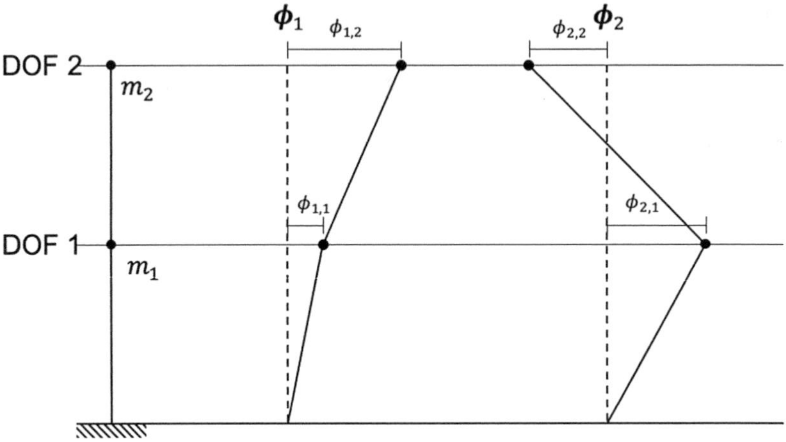

Fig. 4.8 Modal shapes

– *Modal shapes:* $\left(-\omega^2\mathbf{m}+\mathbf{k}\right)\mathbf{\Phi}_i=\mathbf{0}$, i.e.,

$$\begin{bmatrix}-2\omega_i^2+16 & -5\\ -5 & -\omega_i^2+2\end{bmatrix}\begin{bmatrix}\phi_{i,1}\\ \phi_{i,2}\end{bmatrix}=\underline{0}.$$

The first equation with $\phi_{i,1}=1$ gives $\phi_{i,2}=(-2\omega_i^2+16)/5$ so that $\phi_{1,2}=3.0547$ and $\phi_{2,2}=-0.6547$.

MATLAB solution:

```
[u, d] =eig(inv(m)*k)

u =
     0.8366     0.3111
    -0.5478     0.9504
d =
     9.6368          0
          0     0.3632.
```

Step2: Modal coordinates:

$$\mathbf{q}_0=\mathbf{\Psi}^{-1}\mathbf{x}_0=\begin{bmatrix}0.3111 & .8366\\ 0.9504 & -0.5477\end{bmatrix}^{-1}\mathbf{x}_0=\begin{bmatrix}0.5673 & 0.8665\\ 0.9843 & -0.3222\end{bmatrix}\mathbf{x}_0,$$

$$\dot{\mathbf{q}}_0=\mathbf{\Psi}^{-1}\dot{\mathbf{x}}_0=\mathbf{0}\quad\text{so that}\quad q_i(t)=q_{0,i}\cos(\omega_i t),$$

Step3: System displacement:

$$\mathbf{x}(t)=\begin{bmatrix}x_1(t)\\ x_2(t)\end{bmatrix}=\begin{bmatrix}0.3111\\ 0.9506\end{bmatrix}q_{0,1}\cos(\omega_1 t)+\begin{bmatrix}0.8366\\ -0.5478\end{bmatrix}q_{0,2}\cos(\omega_2 t).$$

4.4.5 *Earthquake Engineering*

We present two applications dealing with MDOF systems subjected to support motion, e.g., structural systems subjected to seismic shaking. The first application develops the equations of motions and solutions for this input. The second introduces the concept of design spectrum that is used extensively in Earthquake Engineering. These applications relate to Sect. 4.4.1 on damped systems in forced vibration.

4.4.5.1 Seismic Ground Acceleration

Consider a MDOF structure at a site that is subjected to the seismic ground acceleration $a(t)$. The accelerations of the structural masses have two components. The first component, $\mathbf{m}\,\mathbf{1}\,a(t)$, is generated by the seismic ground acceleration and affects equally all masses. This would be the only component for an infinitely stiff system. The second component, $\mathbf{m}\,\ddot{\mathbf{x}}(t)$, is generated by deformation and differs from mass to mass. The Newton law gives

$$\mathbf{m}\left(\ddot{\mathbf{x}} + \mathbf{1}\,a(t)\right) + \mathbf{c}\,\dot{\mathbf{x}} + \mathbf{k}\,\mathbf{x} = \mathbf{0},$$

where $\mathbf{m}$, $\mathbf{c}$, and $\mathbf{k}$ denote the mass, damping, and stiffness matrices, $\mathbf{x}$ is the displacement vector, and $\mathbf{1}$ denotes the unit column vector in $\mathbb{R}^n$. This gives

$$\mathbf{m}\,\ddot{\mathbf{x}} + \mathbf{c}\,\dot{\mathbf{x}} + \mathbf{k}\,\mathbf{x} = -\mathbf{m}\,\mathbf{1}\,a(t) \tag{4.56}$$

which has the form in Eq. 4.2 with $\mathbf{f}(t) = -\mathbf{m}\,\mathbf{1}\,a(t)$. Accordingly, the method of Sect. 4.4.1 can be applied to find the displacement vector $\mathbf{x}(t)$ for specified initial conditions $(\mathbf{x}_0, \dot{\mathbf{x}}_0)$. We summarize the steps of the analysis for calculating $\mathbf{x}(t)$.

The solution form $\mathbf{x}(t) = \sum_{i=1}^{n} \boldsymbol{\Phi}_i\, q_i(t) = \boldsymbol{\Psi}\,\mathbf{q}(t)$ is that of Eq. 4.19. This solution and Eq. 4.56 give

$$\mathbf{m}\,\boldsymbol{\Psi}\,\ddot{\mathbf{q}}(t) + \mathbf{c}\,\boldsymbol{\Psi}\,\dot{\mathbf{q}}(t) + \mathbf{k}\,\boldsymbol{\Psi}\,\mathbf{q}(t) = -\mathbf{m}\,\mathbf{1}\,a(t),$$

which, by left multiplication with $\boldsymbol{\Psi}^T$, becomes

$$\boldsymbol{\Psi}^T\,\mathbf{m}\,\boldsymbol{\Psi}\,\ddot{\mathbf{q}}(t) + \boldsymbol{\Psi}^T\,\mathbf{c}\,\boldsymbol{\Psi}\,\dot{\mathbf{q}}(t) + \boldsymbol{\Psi}^T\,\mathbf{k}\,\boldsymbol{\Psi}\,\mathbf{q}(t) = -\boldsymbol{\Psi}^T\,\mathbf{m}\,\mathbf{1}\,a(t).$$

The orthogonality in Eq. 4.10 simplifies this equation to

$$\mathrm{diag}\{\tilde{m}_i\}\,\ddot{\mathbf{q}}(t) + \boldsymbol{\Psi}^T\,\mathbf{c}\,\boldsymbol{\Psi}\,\dot{\mathbf{q}}(t) + \mathrm{diag}\{\tilde{k}_i\}\,\mathbf{q}(t) = -\boldsymbol{\Psi}^T\,\mathbf{m}\,\mathbf{1}\,a(t). \tag{4.57}$$

Under the assumption of proportional damping, we have

$$\begin{aligned}
&\operatorname{diag}\{\tilde{m}_i\}\,\ddot{\mathbf{q}}(t) + \operatorname{diag}\{\tilde{c}_i\}\,\dot{\mathbf{q}}(t) + \operatorname{diag}\{\tilde{k}_i\}\,\mathbf{q}(t) = -\boldsymbol{\Psi}^T\,\mathbf{m}\,\mathbf{1}\,a(t) \\
&\Longrightarrow \ddot{q}_i(t) + 2\,\zeta_i\,\omega_i\,\ddot{q}_i(t) + \omega_i^2\,q_i(t) = -\Gamma_i\,a(t), \quad i = 1, \ldots, n,
\end{aligned} \tag{4.58}$$

where $2\,\zeta_i\,\omega_i = \tilde{c}_i/\tilde{m}_i$, $\omega_i^2 = \tilde{k}_i/\tilde{m}_i$, and

$$\Gamma_i = \frac{\left(\boldsymbol{\Psi}^T\,\mathbf{m}\,\mathbf{1}\right)_i}{\tilde{m}_i}, \quad i = 1, \ldots, n, \tag{4.59}$$

denotes the ***modal participation factor*** for mode i. This factor scales the ground acceleration $a(t)$ to $-\Gamma_i\,a(t)$ for the defining equations of the modal coordinates $\{q_i(t)\}$.

The displacement vector $\mathbf{x}(t)$ results from its representation $\mathbf{x}(t) = \sum_{i=1}^n \boldsymbol{\Phi}_i\,q_i(t)$ and the solutions $\{q_i(t)\}$ of Eq. 4.58, which can be obtained by the methods discussed in the first part of the book dealing with SDOF systems. For example, the modal coordinates by the Duhamel's integral are

$$\begin{aligned}
q_i(t) &= e^{-\zeta_i\,\omega_i\,t}\left[q_{i,0}\,\cos\left(\omega_{d,i}\,t\right) + \frac{\dot{q}_{i,0} + \zeta_i\,\omega_i\,q_{i,0}}{\omega_{d,i}}\,\sin\left(\omega_{d,i}\,t\right)\right] \\
&\quad - \Gamma_i \int_0^t \frac{1}{\omega_{d,i}}\,e^{-\zeta_i\,\omega_i\,(t-s)}\,\sin\left(\omega_{d,i}\,(t-s)\right)a(s)\,ds
\end{aligned} \tag{4.60}$$

so that the system displacement has the form

$$\begin{aligned}
\mathbf{x}(t) &= \sum_{i=1}^n \boldsymbol{\Phi}_i\,e^{-\zeta_i\,\omega_i\,t}\left[q_{i,0}\,\cos\left(\omega_{d,i}\,t\right) + \frac{\dot{q}_{i,0} + \zeta_i\,\omega_i\,q_{i,0}}{\omega_{d,i}}\,\sin\left(\omega_{d,i}\,t\right)\right] \\
&\quad - \sum_{i=1}^n \boldsymbol{\Phi}_i\,\Gamma_i \int_0^t \frac{1}{\omega_{d,i}}\,e^{-\zeta_i\,\omega_i\,(t-s)}\,\sin\left(\omega_{d,i}\,(t-s)\right)a(s)\,ds.
\end{aligned} \tag{4.61}$$

The vector of seismic forces at the structural degrees of freedom has the expression

$$\mathbf{F}_{\text{seismic}}(t) = \mathbf{k}\,\mathbf{x}(t) = \sum_{i=1}^n \mathbf{k}\,\boldsymbol{\Phi}_i\,q_i(t) = \sum_{i=1}^n \mathbf{m}\,\boldsymbol{\Phi}_i\,\omega_i^2\,q_i(t). \tag{4.62}$$

The above unit impulse response function,

$$h_i(t-s) = \frac{1}{\omega_{d,i}}\,e^{-\zeta_i\,\omega_i\,(t-s)}\,\sin\left(\omega_{d,i}\,(t-s)\right), \quad t \geq s, \tag{4.63}$$

differs slightly from that in Eq. 2.30 for SDOF systems with displacement functions $x(t)$ defined by the differential equation $\ddot{x} + 2\,\zeta\,\omega\,\dot{x} + \omega^2\,x = f(t)/m$ since the analogue of $f(t)$ in this equation is $-\big(\Gamma_i\,a(t)\big)\,\tilde{m}_i$ for the differential equation of $q_i(t)$ so that its forced vibration component is (see Eq. 2.30)

$$q_{p,i}(t) = \int_0^t \frac{1}{\tilde{m}_i\,\omega_{d,i}}\,e^{-\zeta_i\,\omega_i\,(t-s)}\,\sin\big(\omega_{d,i}\,(t-s)\big)\,\big(-\Gamma_i\,a(s)\,\tilde{m}_i\big)\,ds$$

$$= -\Gamma_i \int_0^t \frac{1}{\omega_{d,i}}\,e^{-\zeta_i\,\omega_i\,(t-s)}\,\sin\big(\omega_{d,i}\,(t-s)\big)\,a(s)\,ds, \qquad (4.64)$$

which gives the unit impulse response function in Eq. 4.63.

4.4.5.2 Design Response Spectrum

Recall that response spectra are maxima of responses of SDOF systems, e.g., the displacement response spectrum $S_d(\omega, \zeta)$ is the maxima of the absolute values of displacements of a family of SDOF systems indexed by their natural frequency ω and damping ratio ζ that are subjected to the same ground acceleration $a(t)$. Accordingly, the maxima of the modal responses given by Eq. 4.60 or other forms of the solution of Eq. 4.58 are $R_i = \max_t\{|q_i(t)|\} = \Gamma_i\,S_d(\omega_i, \zeta_i)$, $i = 1, \ldots, n$, since we deal with linear differential equations.

The modal spectral displacements $\{R_i\}$ are well defined. However, system response maxima cannot be obtained exactly from $\{R_i\}$ since maxima of different modal coordinates occur at different times. To overcome this impasse, it is common to use ***modal combination rules***, i.e., formulas that map modal response maxima into system response maxima. The rules are based on the qualitative arguments, e.g., it is assumed that modal responses with closely spaced frequencies are in phase so that their extremes occur simultaneously. The meaning of closely spaced is not defined. Moreover, modal frequency closeness is affected by modal damping ratio and the frequency content of $a(t)$. The estimation of structural response maxima, e.g., the displacement maxima, by modal combination rules involves two steps.

- *Step 1*: Find $S_d(\omega_i, \zeta_i)$ from plots of response spectra for the ground acceleration $a(t)$ of interest and calculate modal response maxima from $R_i = \max_t\{|q_i(t)|\} = \Gamma_i\,S_d(\omega_i, \zeta_i)$.
- *Step 2*: Estimate structural response maxima from modal response maxima $\{R_i\}$ by modal combination rules. Here are two of the most popular modal combination rules.
  - SRSS (Square root of the sum of the square of modal response maxima): $R = \big(\sum_{i=1}^n R_i^2\big)^{1/2}$ if modal frequencies are not closely spaced.
  - ABS (Sum of modal response maxima): $R = \sum_{i=1}^n R_i$, if modal frequencies are closely spaced.

Note that (1) ABS-based responses always exceed structural responses since maxima of modal responses do not occur simultaneously, (2) ABS- and SRSS-based responses differ significantly if modal response maxima are similar, e.g., they are $n\,R_1$ and $\sqrt{n}\,R_1$ if $R_i = R_1,\ i \geq 2$, and (3) ABS- and SRSS-based responses are similar if one of the modal responses is dominant, e.g., they are approximately equal to R_1 if $R_1 \gg R_i,\ i \geq 2$.

Example 4.6 Suppose the 2-DOF system with mass and stiffness matrices given in Example 4.3, and a damping matrix $\mathbf{c} = \alpha\,\mathbf{m} + \beta\,\mathbf{k}$, where $\alpha = 0.03$ and $\beta = 0.05$, is subjected to the El Centro ground motion. The modal periods and damping ratios are $T_1 = 10.425$ s, $T_2 = 2.0240$ s, $\zeta_1 = 0.04$, and $\zeta_2 = 0.0824$. The response spectral values are $S_d(\omega_1, \zeta_1) = 16$ in and $S_d(\omega_2, \zeta_2) = 8$ in. They can be obtained from the response spectrum of this seismic event shown in Fig. 2.17. The modal participation factors are $\Gamma_1 = 1.4338$ and $\Gamma_2 = 0.6621$ so that the maxima of modal displacements are $R_1 = 22.94$ in and $R_2 = 5.30$ in. The structural responses $\{\mathbf{\Phi}_i\ R_i\}$ in the two modes are

$$\mathbf{\Phi}_1\,R_1 = \begin{bmatrix} 7.14 \\ 21.86 \end{bmatrix} \text{ in} \quad \text{and } \mathbf{\Phi}_2\,R_2 = \begin{bmatrix} 4.43 \\ -2.90 \end{bmatrix} \text{ in.}$$

The ABS- and SRSS-estimates of the drift are

$$\begin{aligned} &\text{ABS:}|21.80| + |-2.90| = 24.70\text{in} \\ &\text{SRSS:}\left(21.80^2 + (-2.90)^2\right)^{1/2} = 21.99\text{in.} \end{aligned}$$

The ABS- and SRSS-estimates of the inter-story displacement are

$$\begin{aligned} &\text{ABS: } |21,80 - 7.14| + |4,43 - (-2.90)| = 21.99\text{in,} \\ &\text{SRSS: } \left((21,80 - 7.14)^2 + (4,43 - (-2.90)^2)\right)^{1/2} = 16.39\text{in.} \end{aligned}$$

Note that drift estimates by the ABS and SRSS rules are similar. In contrast, the inter-story estimates by these rules differ significantly. Since the modal periods T_1 and T_2 are not closely spaced (a qualitative statement!), it is likely that the SRRS rule would be used for design. As expected, the ABS-based estimates are larger than the SRSS-based estimates.

4.4.6 Experimental Determination of Modal Frequencies

Consider a MDOF system with n degrees of freedom, mass matrix $\mathbf{m}$, and stiffness matrix $\mathbf{k}$, which is placed on a shake table oscillating with the acceleration $\sin(\nu\,t)$. The frequency ν of the forcing function is increased slowly, and the resulting structural responses are monitored. We show that this sequence of experiments can be used to estimate the modal frequencies $\{\omega_i\}$. This application relates to Sect. 4.4.3 on the undamped systems in forced vibration.

The equation of motion of the system is (see Eq. 4.56)

$$\mathbf{m}\,\ddot{\mathbf{x}} + \mathbf{k}\,\mathbf{x} = -\mathbf{m}\,\mathbf{1}\,\sin(\nu\,t), \tag{4.65}$$

where **1** denotes the n-dimensional unit column vector. Considerations as those used to obtain Eq. 4.25 give

$$\ddot{q}_i(t) + \omega_i^2\, q_i(t) = -\frac{\left(\boldsymbol{\Psi}^T\,\mathbf{m}\,\mathbf{1}\right)_i}{\tilde{m}_i}\,\sin(\nu\,t), \quad i = 1, \ldots, n, \tag{4.66}$$

so that the particular solution is (see Sect. 2.4.5 on SDOF systems)

$$q_i(t) = \frac{\beta_i/\tilde{k}_i}{1-(\nu/\omega_i)^2}\left[\sin(\nu\,t) - \frac{\nu}{\omega_i}\,\sin(\omega_i\,t)\right], \quad \nu \neq \omega_i, \quad i = 1, \ldots, n, \tag{4.67}$$

where $\beta_i = \left(\boldsymbol{\Psi}^T\,\mathbf{m}\,\mathbf{1}\right)_i$. The displacement vector has the expression

$$\mathbf{x}(t) = \sum_{i=1}^{n} \boldsymbol{\Phi}_i\, q_i(t) = \sum_{i=1}^{n} \boldsymbol{\Phi}_i\,\frac{\beta_i/\tilde{k}_i}{1-(\nu/\omega_i)^2}\left[\sin(\nu\,t) - \frac{\nu}{\omega_i}\,\sin(\omega_i\,t)\right]. \tag{4.68}$$

The above expression of the displacement vector $\mathbf{x}(t)$ is used to explain the behavior of this vector as the forcing frequency is increased slowly from zero and estimate the modal frequencies.

1. Suppose that $\nu \ll \omega_i$, $i = 1, \ldots, n$. Then, the dynamic amplification factors $\text{DAF}_i = 1/\left(1-(\nu/\omega_i)^2\right) \simeq 1$, $i = 1, \ldots, n$, so that there is no dynamic amplification. The system response is quasi-static.
2. Increase ν so that it is in a small vicinity of ω_1. Then, $\text{DAF}_1 \gg \text{DAF}_i$, $i \neq 1$, so that $\mathbf{x}(t) \simeq \boldsymbol{\Phi}_1\, q_1(t)$. The system motion is dominated by mode 1 and $\omega_1 \simeq \nu$.
3. Increase further ν so that it is in a small vicinity of ω_2. Then, $\text{DAF}_2 \gg \text{DAF}_i$, $i \neq 2$, so that $\mathbf{x}(t) \simeq \boldsymbol{\Phi}_2\, q_2(t)$. The system motion is dominated by mode 2 and $\omega_2 \simeq \nu$.
4. The values of ν corresponding to significant increases in the system response are approximations of the modal frequencies. The corresponding deformations are approximate modal shapes.

4.4.7 Torsional Vibration

Generally, structural and mechanical systems are three-dimensional and their motions involve translations and rotations rather than just translation in two-dimensional spaces. For simplicity, we consider the single story structure in Fig. 4.9 and assume that (1) the entire mass is concentrated at floor level, (2) the floor is infinitely stiff in its own plane, (3) the deformation is small so that the motion of the

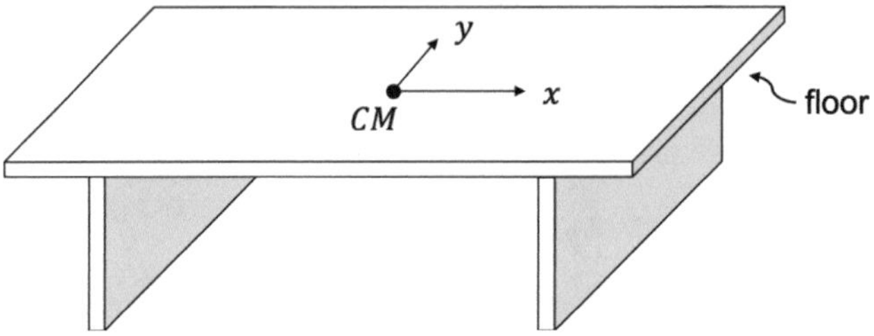

Fig. 4.9 Single story structure (3D view)

floor, a rigid body in $\mathbb{R}^3$, can be described by two translations and a rotation, and (4) the vertical structures are planar with no stiffness outside their planes.

We construct the structural stiffness and mass matrices, establish the equation of motion, and illustrate the implementation of these developments for a simple structure. It is shown that our tools developed so far are adequate to characterize the vibration of this three-dimensional system. This application relates to Sect. 4.4.4 on the undamped systems in free vibration.

Stiffness matrix We use the physical arguments to construct the stiffness matrix of the structure in Fig. 4.9, i.e., we impose a displacement u along the x-coordinate, a displacement v along the y-coordinate, and a rotation θ about the center of mass (CM) of this system and record the induced forces.

The displacement in the plane of the ith vertical supporting structure induced by the displacements (u, v, θ) of the floor is

$$\delta_i = u\,\cos(\alpha_i) + v\,\sin(\alpha_i) - \theta\,d_i, \tag{4.69}$$

with the notations and sign convention of Fig. 4.10. The force induced in this vertical structure has the expression

$$F_i = k_i\,\delta_i = \big(k_i\,\cos(\alpha_i)\big)\,u + \big(k_i\,\sin(\alpha_i)\big)\,v - \big(k_i\,d_i\big)\,\theta, \tag{4.70}$$

where k_i denotes its in-plane stiffness.

The total forces induced by the floor displacements (u, v, θ) result by summing the projections of the forces $\{F_i\}$ on the x- and y-coordinates and by summing the moments of these forces with respect to the center of mass, i.e.,

$$\begin{aligned} F_x &= \sum_i F_i\,\cos(\alpha_i) \\ &= \big(\sum_i k_i\,\cos^2(\alpha_i)\big)\,u + \big(\sum_i k_i\,\cos(\alpha_i)\,\sin(\alpha_i)\big)\,v - \big(\sum_i k_i\,d_i\,\cos(\alpha_i)\big)\,\theta \end{aligned}$$

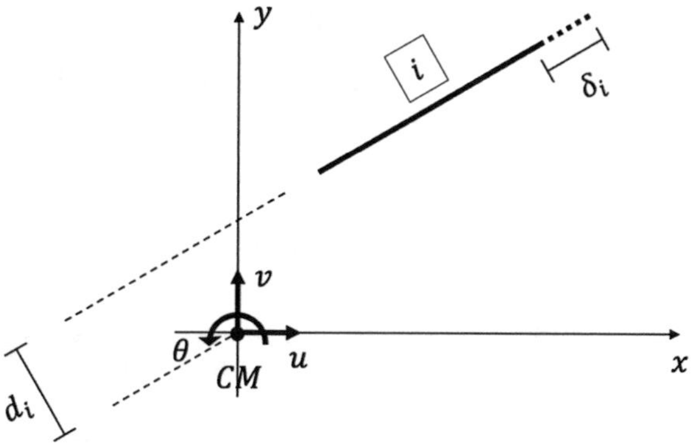

Fig. 4.10 Top view of a single story structure and positive sign convention. Structure i is at angle α relative to the x-axis

$$
\begin{aligned}
F_y &= \sum_i F_i \sin(\alpha_i) \\
&= \big(\sum_i k_i \cos(\alpha_i) \sin(\alpha_i)\big)\, u + \big(\sum_i k_i \sin^2(\alpha_i)\big)\, v - \big(\sum_i k_i\, d_i \sin(\alpha_i)\big)\, \theta \\
T &= -\sum_i F_i\, d_i \\
&= -\big(\sum_i k_i\, d_i \cos(\alpha_i)\big)\, u - \big(\sum_i k_i\, d_i \sin(\alpha_i)\big)\, v + \big(\sum_i k_i\, d_i^2\big)\, \theta
\end{aligned}
\tag{4.71}
$$

or, in matrix form,

$$
\begin{bmatrix} F_x \\ F_y \\ T \end{bmatrix} = \begin{bmatrix} \sum_i k_i \cos^2(\alpha_i) & \sum_i k_i \cos(\alpha_i) \sin(\alpha_i) & -\sum_i k_i\, d_i \cos(\alpha_i) \\ \sum_i k_i \cos(\alpha_i) \sin(\alpha_i) & \sum_i k_i \sin^2(\alpha_i) & -\sum_i k_i\, d_i \sin(\alpha_i) \\ -\sum_i k_i\, d_i \cos(\alpha_i) & -\sum_i k_i\, d_i \sin(\alpha_i) & \sum_i k_i\, d_i^2 \end{bmatrix}
\times \begin{bmatrix} u \\ v \\ \theta \end{bmatrix}, \tag{4.72}
$$

so that the stiffness matrix is given by

$$
\mathbf{k} = \begin{bmatrix} \sum_i k_i \cos^2(\alpha_i) & \sum_i k_i \cos(\alpha_i) \sin(\alpha_i) & -\sum_i k_i\, d_i \cos(\alpha_i) \\ \sum_i k_i \cos(\alpha_i) \sin(\alpha_i) & \sum_i k_i \sin^2(\alpha_i) & -\sum_i k_i\, d_i \sin(\alpha_i) \\ -\sum_i k_i\, d_i \cos(\alpha_i) & -\sum_i k_i\, d_i \sin(\alpha_i) & \sum_i k_i\, d_i^2 \end{bmatrix}. \tag{4.73}
$$

Mass matrix Denote by $\bar{m}$ the mass per unit floor area so that the total floor mass (structural mass under our assumptions) is $m = \int_{\text{floor area}} \bar{m}\, dx\, dy$. The inertia forces corresponding to translations in the x- and y-directions are $-m\, \ddot{u}$ and $-m\, \ddot{v}$. The inertia corresponding to rotation is $-I_p\, \bar{m}\, \ddot{\theta}$, where $I_p = \int_{\text{floor area}} r^2\, \bar{m}\, dx\, dy$ denotes the polar moment of inertia so that the mass matrix is

$$\mathbf{m} = \begin{bmatrix} m & 0 & 0 \\ 0 & m & 0 \\ 0 & 0 & I_p \end{bmatrix}. \tag{4.74}$$

The polar moment of inertia can be calculated from

$$\begin{aligned} I_p &= \int_{\text{floor area}} r^2\, \bar{m}\, dA = \bar{m} \int_{\text{floor area}} (x^2 + y^2)\, dA \\ &= \bar{m} \int x^2\, dA + \bar{m} \int y^2\, dA = \bar{m}\,\big(b\, d^3/12 + b^3\, d/12\big) = m\,(b^2 + d^2)/12, \end{aligned}$$

under the assumptions that $\bar{m}$ is constant and the floor is rectangular with sides b and d.

Equation of motion Consider the free vibration of the single-floor structure in Fig. 4.9 under initial conditions (u_0, v_0, θ_0). The equation of motion is

$$\mathbf{m}\ddot{\mathbf{x}} + \mathbf{k}\,\mathbf{x} = \mathbf{0}, \tag{4.75}$$

where $\mathbf{x} = [u\ v\ \theta]^T$ is the displacement vector. We use the previous methods to find the solution of Eq. 4.75, i.e., we calculate the modal shapes and frequencies for the mass and stiffness matrices in Eqs. 4.73 and 4.75, represent the displacement vector by $\mathbf{x}(t) = \sum_{i=1}^{3} \mathbf{\Phi}_i\, q_i(t)$, and use orthogonality and Eq. 4.75 to find equations for modal coordinates.

Example 4.7 Consider the single story in Fig. 4.11 supported by three vertical structures with in-plane stiffnesses $k_A = 75$ kip/ft and $k_B = k_C = 40$ kip/ft and dimensions $b = 30$ ft, $d = 20$ ft, and $e = 1.5$ ft. The floor weight per unit area is $w = 100$ lb/ft^2. The stiffness matrix is

$$\mathbf{k} = \begin{bmatrix} k_B + k_C & 0 & (k_C - k_B)\,(d/2) \\ 0 & k_A & k_A\, e \\ (k_C - k_B)\,(d/2) & k_A\, e & (k_C + k_B)\,(d/2)^2 + k_A\, e^2 \end{bmatrix} = \begin{bmatrix} 80 & 0 & 0 \\ 0 & 75 & 112.5 \\ 0 & 112.5 & 8168.75 \end{bmatrix}.$$

It can be obtained from its expression in Eq. 4.73 or by direct considerations. For example, the forces induced in the vertical supporting structures B and C by the displacement $u = 1$ are k_B and k_C. They act in the positive direction of the x-axis. There is no force in A. This unit displacement induces the force $k_B + k_C$ in the x-direction, no force in the the y-direction, and the torque $k_B\,(d/2) - k_C\,(d/2)$ under the sign convention of Fig. 4.10. Accordingly, $\big((k_B + k_C, 0, (k_B - k_C)\, d/2\big)$ are the entries of the first column of the stiffness matrix $\mathbf{k}$. The reader is encouraged to find the other entries of $\mathbf{k}$ by using similar arguments.

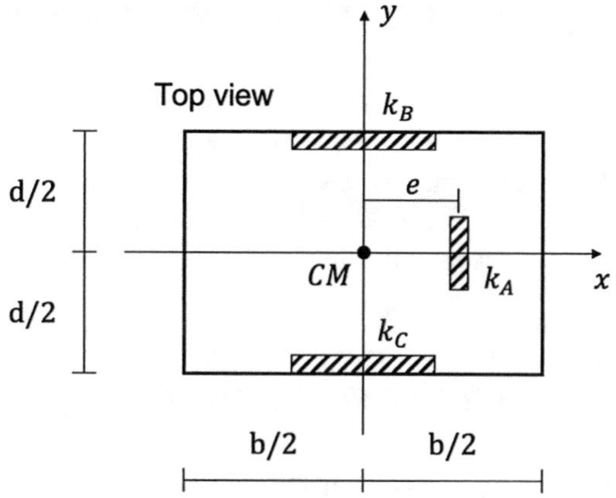

Fig. 4.11 Top view of a single story structure

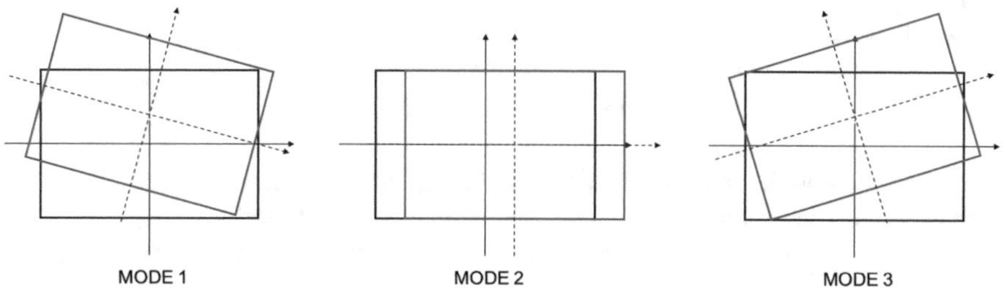

Fig. 4.12 Modal shapes of the structure in Fig. 4.11

The mass matrix results from Eq. 4.74 and is

$$\mathbf{m} = \begin{bmatrix} 1.863 & 0 & 0 \\ 0 & 1.863 & 0 \\ 0 & 0 & 201.863 \end{bmatrix}.$$

The MATLAB function $[\mathbf{u}, \mathbf{d}] = \text{eig}(\mathbf{m}^{-1}\,\mathbf{k})$ gives the following modal shapes and frequencies:

$$\mathbf{u} = \begin{bmatrix} 1 & 0 & 0 \\ 0 & 0.9956 & 0.9953 \\ 0 & -0.0939 & 0.0974 \end{bmatrix} \text{ and } \mathbf{d} = \begin{bmatrix} 6.5530 & 0 & 0 \\ 0 & 5.8788 & 0 \\ 0 & 0 & 6.7944 \end{bmatrix},$$

which are visualized in Fig. 4.12 with the sign convention of Fig. 4.10.

4.5 Time Domain Analysis: Non-proportional Damping

We consider non-proportional damping matrices $\mathbf{c}$, i.e., damping matrices for which $\boldsymbol{\Psi}^T\,\mathbf{c}\,\boldsymbol{\Psi}$ is not diagonal, where $\boldsymbol{\Psi}$ denotes the matrix of classical modes of vibration. This means that the previous method for solving the equation of motion cannot be used since Eqs. 4.22 and 4.23 remain coupled. We develop an alternative method that is conceptually similar to that based on classical modes of vibration.

4.5.1 *State-Space Representation*

Consider a damped MDOF system subjected to a forcing function whose displacement $\mathbf{x}(t)$ is the solution of Eq. 4.2 with the initial conditions $(\mathbf{x}_0, \dot{\mathbf{x}}_0)$. This equation, written here for convenience with a different number, has the form

$$\mathbf{m}\,\ddot{\mathbf{x}} + \mathbf{c}\,\dot{\mathbf{x}} + \mathbf{k}\,\mathbf{x} = \mathbf{f}(t), \tag{4.76}$$

where $\mathbf{m}$, $\mathbf{c}$, and $\mathbf{k}$ denote the mass, damping, and stiffness matrices, $\mathbf{x}$ is the displacement vector, and $\mathbf{f}(t)$ is an n-dimensional forcing function.

The ***state-space*** representation of the equation of motion has the form

$$\dot{\mathbf{z}}(t) = \mathbf{a}\,\mathbf{z}(t) + \mathbf{b}\,\mathbf{f}(t), \tag{4.77}$$

where

$$\mathbf{z}(t) = \begin{bmatrix} \mathbf{x}(t) \\ \dot{\mathbf{x}}(t) \end{bmatrix}, \quad \mathbf{a} = \begin{bmatrix} \mathbf{0} & \mathbf{I} \\ -\mathbf{m}^{-1}\,\mathbf{k} & -\mathbf{m}^{-1}\,\mathbf{c} \end{bmatrix} \quad \text{and} \quad \mathbf{b} = \begin{bmatrix} \mathbf{0} \\ \mathbf{m}^{-1} \end{bmatrix}. \tag{4.78}$$

We note that (1) the first n and the last n components of the $2\,n$-dimensional vector $\mathbf{z}(t)$ are the displacement and velocity vectors $\mathbf{x}(t)$ and $\dot{\mathbf{x}}(t)$, (2) the first n equations in Eq. 4.78 state that the derivative of $\mathbf{x}(t)$ is $\dot{\mathbf{x}}(t)$, (3) the last n equations result from Eq. 4.76 by left multiplication with $\mathbf{m}^{-1}$, (4) the state vector $\mathbf{z}(t)$ is equal to $\mathbf{z}_0 = [\mathbf{x}_0^T\ \dot{\mathbf{x}}_0^T]^T$ at the initial time $t = 0$, and (5) the $(2\,n, 2\,n)$-matrix $\mathbf{a}$ is not symmetric.

4.5.2 *Eigenvalues and Right/Left Eigenvectors*

We proceed as in Sect. 4.3 to construct an eigenvalue problem for the state-space equation of motion. Consider the trial solution

$$\mathbf{z}(t) = \mathbf{w}\,e^{\lambda\,t}, \tag{4.79}$$

for the homogeneous version of Eq. 4.77, i.e., $\mathbf{f}(t) = \mathbf{0}$, where the $2n$-dimensional column vector $\mathbf{w}$ and the scalar λ are unknown and need to be determined. We show that the solutions of this type can be used to construct sets of $2n$-dimensional vectors that span $\mathbb{R}^{2n}$ so that $\mathbf{z}(t)$ can be represented in the coordinate systems defined by these sets of vectors.

The expressions of $\mathbf{z}(t)$ in Eq. 4.79 and Eq. 4.77 with $\mathbf{f}(t) = \mathbf{0}$ give

$$\mathbf{w}\,\lambda\, e^{\lambda t} = \mathbf{a}\,\mathbf{w}\, e^{\lambda t} \quad \text{or, equivalently,} \quad \left(\mathbf{a} - \lambda\,\mathbf{I}\right)\mathbf{w}\, e^{\lambda t} = \mathbf{0},$$

which implies

$$\left(\mathbf{a} - \lambda\,\mathbf{I}\right)\mathbf{w} = \mathbf{0}, \tag{4.80}$$

since $\left(\mathbf{a} - \lambda\,\mathbf{I}\right)\mathbf{w}\, e^{\lambda t} = \mathbf{0}$ must hold at all times and $\exp(\lambda\, t)$ is not zero for bounded λ. The condition of Eq. 4.80 defines a linear homogeneous system of equations for $\mathbf{w}$ that admits the trivial solution if $\det\left(\mathbf{a} - \lambda\,\mathbf{I}\right) \neq 0$ and non-trivial solutions, in addition to the trivial solution, if $\det\left(\mathbf{a} - \lambda\,\mathbf{I}\right) = 0$. Since the trivial solution is not possible for non-zero initial conditions ($\mathbf{z}(t) = \mathbf{0}$ at all times if $\mathbf{w} = \mathbf{0}$), we require

$$\det\left(\mathbf{a} - \lambda\,\mathbf{I}\right) = 0. \tag{4.81}$$

For simplicity, we assume that the $2n$ solutions $\{\lambda_i\}$, $i = 1, \ldots, 2n$, of these equations, i.e., the ***eigenvalues*** of the nonsymmetric matrix $\mathbf{a}$, are distinct.

As for classical modes of vibration, the solutions of the above eigenvalue problem involve the following two steps.

- *Step 1. Eigenvalues*: The roots $\lambda_1, \ldots, \lambda_{2n}$ of the $2n$-degree polynomial $\det\left(\mathbf{a} - \lambda\,\mathbf{I}\right)$ in λ, i.e., the solutions of Eq. 4.81, are the ***eigenvalues*** of matrix $\mathbf{a}$. Since $\det\left(\mathbf{a} - \lambda\,\mathbf{I}\right) = 0$ and $\det\left(\mathbf{a} - \lambda\,\mathbf{I}\right)^T = \det\left(\mathbf{a}^T - \lambda\,\mathbf{I}\right) = 0$ have the same roots (the determinant of a matrix coincides with the determinant of its transposed), the system has a single set of $2n$ eigenvalues. Generally, the eigenvalues are complex-valued as $\mathbf{a}$ is not symmetric.
- *Step 2. Eigenvectors*: There are two sets of eigenvectors, ***right eigenvectors*** and ***left eigenvectors***, which correspond to the non-trivial solutions of $\left(\mathbf{a}-\lambda\,\mathbf{I}\right)\mathbf{w} = \mathbf{0}$, and $\left(\mathbf{a}^T - \lambda\,\mathbf{I}\right)\mathbf{w} = \mathbf{0}$, i.e.,

$$\begin{aligned} \mathbf{a}\,\mathbf{u}_i &= \lambda_i\,\mathbf{u}_i \quad \text{(Right eigenvectors)} \\ \mathbf{a}^T\,\mathbf{v}_i &= \lambda_i\,\mathbf{v}_i \quad \text{(Left eigenvectors)}. \end{aligned} \tag{4.82}$$

We use the notations $\mathbf{u}$ and $\mathbf{v}$ for the right and left eigenvectors of matrix $\mathbf{a}$.

4.5.3 Properties of Eigenvalues and Right/Left Eigenvectors

We limit our discussion to properties of the eigenvalues and the right/left eigenvectors, which is relevant to our discussion. As stated, we assume distinct eigenvalues for simplicity.

1. **The right and left eigenvectors corresponding to distinct eigenvalues are orthogonal** in the sense that

$$\mathbf{v}_j^T\,\mathbf{u}_i = 0 \quad \text{and} \quad \mathbf{v}_j^T\,\mathbf{a}\,\mathbf{u}_i = 0, \quad i \neq j. \tag{4.83}$$

Proof Consider two distinct eigenvalues $\lambda_i \neq \lambda_j$, $i \neq j$. The first set of equations in Eq. 4.82 becomes $\mathbf{v}_j^T\,\mathbf{a}\,\mathbf{u}_i = \lambda_i\,\mathbf{v}_j^T\,\mathbf{u}_i$ by left multiplication with $\mathbf{v}_j^T$. The transposed of the second set of equations in Eq. 4.82 for the left vector j becomes $\mathbf{v}_j^T\,\mathbf{a}\,\mathbf{u}_i = \lambda_j\,\mathbf{v}_j^T\,\mathbf{u}_i$ by right multiplication with $\mathbf{u}_i$. The left sides of the resulting two equations coincide so that their right sides must coincide, i.e., $\lambda_i\,\mathbf{v}_j^T\,\mathbf{u}_i = \lambda_j\,\mathbf{v}_j^T\,\mathbf{u}_i$ or $(\lambda_i - \lambda_j)\,\mathbf{v}_j^T\,\mathbf{u}_i = 0$. Since $\lambda_i \neq \lambda_j$ by assumption, we have $\mathbf{v}_j^T\,\mathbf{u}_i = 0$. The above equalities also give $\mathbf{v}_j^T\,\mathbf{a}\,\mathbf{u}_i = 0$ for $i \neq j$.

The equality $\mathbf{v}_j^T\,\mathbf{a}\,\mathbf{u}_i = \lambda_i\,\mathbf{v}_j^T\,\mathbf{u}_i$ in the above arguments written for $i = j$ gives

$$\lambda_i = \frac{\mathbf{v}_i^T\,\mathbf{a}\,\mathbf{u}_i}{\mathbf{v}_i^T\,\mathbf{u}_i}, \quad i = 1, \ldots, 2n. \tag{4.84}$$

2. **The matrix form of the orthogonality condition** is

$$\mathbf{v}^T\,\mathbf{u} = \text{diag}(\mathbf{v}_i^T\,\mathbf{u}_i) \quad \text{and} \quad \mathbf{v}^T\,\mathbf{a}\,\mathbf{u} = \text{diag}(\mathbf{v}_i^T\,\mathbf{a}\,\mathbf{u}_i), \tag{4.85}$$

where $\mathbf{u} = [\mathbf{u}_1\,\mathbf{u}_2 \ldots \mathbf{u}_{2n}]$ and $\mathbf{v} = [\mathbf{v}_1\,\mathbf{v}_2 \ldots \mathbf{v}_{2n}]$ are $(2n, 2n)$-matrices whose columns are right and left eigenvectors.

3. **If λ is an eigenvalue, its complex conjugate λ^* is also an eigenvalue.**

Proof The eigenvalue λ is a root of the polynomial $\sum_{i=0}^{2n} a_i\,\lambda^i$ with real-valued coefficients, i.e., it is a solution of $\sum_{i=0}^{2n} a_i\,\lambda^i = 0$. Since the complex conjugate is commutative with integer powers, i.e., $(\lambda^i)^* = (\lambda^*)^i$, the complex conjugate of $\sum_{i=0}^{2n} a_i\,\lambda^i = 0$ is $\sum_{i=0}^{2n} a_i\,(\lambda^*)^i = 0$ so that, if λ is an eigenvalue, so is λ^*.

4. **If u is the right eigenvector corresponding to an eigenvalue λ, then $\mathbf{u}^*$ is the right eigenvector of λ^*.**

Proof Under the assumption that $\mathbf{u}$ is the right eigenvector of an eigenvalue λ, we have $\mathbf{a}\,\mathbf{u} = \lambda\,\mathbf{u}$. The complex conjugate of this equality, $\mathbf{a}\,\mathbf{u}^* = \lambda^*\,\mathbf{u}^*$, shows that $\mathbf{u}^*$ is a right eigenvector corresponding to the eigenvalue λ^*.

Example 4.8 The eigenvalues and the right/left eigenvectors of the (2,2)-matrix

$\mathbf{a} = \begin{bmatrix} 0 & 1 \\ -36 & -0.6 \end{bmatrix}$ are the solutions of $\det\left(\mathbf{a} - \lambda\,\mathbf{I}\right) = \det\left(\mathbf{a}^T - \lambda\,\mathbf{I}\right) = \lambda^2 + 0.6\,\lambda + 36 = 0.$

They can be found by direct calculations from the above equation. However, it is convenient to use MATLAB that provides an efficient solution through the functions

$$\begin{aligned} [\mathbf{u}, \mathbf{d}] &= \mathrm{eig}(\mathbf{a}) \quad \text{and} \\ [\mathbf{v}, \mathbf{d}] &= \mathrm{eig}(\mathbf{a}'), \end{aligned} \tag{4.86}$$

where the columns of the (2,2)-matrices **u** and **v** are the right and left eigenvectors, the non-zero entries of **d** are the eigenvalues of **a** or, equivalently, $\mathbf{a}^T$. Recall that the MATLAB notation for $\mathbf{a}^T$ is $\mathbf{a}'$. The outputs of Eq. 4.86 are

$$\mathbf{u} = \begin{bmatrix} -0.0082 - 0.1642\,i & -0.0082 + 0.1642\,i \\ 0.9864 + 0.0000\,i & 0.9864 + 0.0000i \end{bmatrix},$$

$$\mathbf{v} = \begin{bmatrix} -0.9864 + 0.0000\,i & -0.9864 + 0.0000\,i \\ -0.0082 + 0.1642\,i & -0.0082 - 0.1642\,i \end{bmatrix},$$

$$\mathbf{d} = \begin{bmatrix} -0.3000 + 5.9925\,i & 0.0000 + 0.0000\,i \\ 0.0000 + 0.0000\,i & -0.3000 - 5.9925\,i \end{bmatrix},$$

where $i = \sqrt{-1}$ denotes the imaginary unit. Note that the eigenvalues are the complex conjugates and so are their right/left eigenvectors and that the eigenvectors are orthogonal in the sense of Eqs 4.83 and 4.85. The orthogonality condition can be checked in MATLAB by writing $\mathbf{v}.' * \mathbf{u}$ and $\mathbf{v}.' * \mathbf{a} * \mathbf{u}$, where the notation $\mathbf{v}.'$ means matrix transposition and has to be used for complex-valued matrices, such as matrix **v**. For our case, these conditions are

$$\mathbf{v}.' * \mathbf{u} = \begin{bmatrix} 0.0000 + 0.323\,i & 0.0000 - 0.0000\,i \\ 0.0000 + 0.0000\,i & 0.0000 - 0.3239\,i \end{bmatrix} \quad \text{and}$$

$$\mathbf{v}.' * \mathbf{a} * \mathbf{u} = \begin{bmatrix} -1.9411 - 0.0972\,i & 0.0000 - 0.0000\,i \\ 0.0000 + 0.0000\,i & -1.9411 + 0.0972\,i \end{bmatrix}.$$

The eigenvalues are the non-zero entries $\lambda_1 = -0.3 + 5.9925\,i$ and $\lambda_2 = -0.3 - 5.9925\,i$ of **d**. The columns of **u** and **v** are the right and left eigenvectors. For examples, the first right and left eigenvectors are the first columns of **u** and **v**.

4.5.4 Forced Vibration

We have seen that, if the eigenvalues are distinct, the right and left eigenvectors provide basis in $\mathbb{R}^{2n}$ since they are linearly independent (see Sect. 3.2.2). Accordingly, the $2n$-dimensional vector $\mathbf{z}(t)$ can be represented at any time t by its projections on the right or left eigenvectors, e.g.,

$$\mathbf{z}(t) = \sum_{i=1}^{2n} \mathbf{u}_i\, q_i(t) = \mathbf{u}\,\mathbf{q}(t) \tag{4.87}$$

with the notation $\mathbf{q}(t) = [q_1(t)\; q_2(t) \ldots q_{2n}(t)]^T$. As previously, we refer to the components of $\mathbf{q}(t)$ as ***modal coordinates***. If the eigenvalues are not distinct, the set of right/left eigenvectors and their corresponding generalized eigenvectors have to be used to represent the state vector $\mathbf{z}(t)$.

The representations of $\mathbf{z}(t)$ in Eq. 4.87 and Eq. 4.77 give

$$\begin{aligned}
&\mathbf{u}\,\dot{\mathbf{q}}(t) = \mathbf{a}\,\mathbf{u}\,\mathbf{q}(t) + \mathbf{b}\,\mathbf{f}(t) \quad \text{(which becomes)}\\
&\mathbf{v}^T\,\mathbf{u}\,\dot{\mathbf{q}}(t) = \mathbf{v}^T\,\mathbf{a}\,\mathbf{u}\,\mathbf{q}(t) + \mathbf{v}^T\,\mathbf{b}\,\mathbf{f}(t) \quad \text{(by left multiplication with } \mathbf{v}^T)\\
&\mathbf{v}_i^T\,\mathbf{u}_i\,\dot{q}_i(t) = \mathbf{v}_i^T\,\mathbf{a}\,\mathbf{u}_i\,q_i(t) + \left(\mathbf{v}^T\,\mathbf{b}\,\mathbf{f}(t)\right)_i, \quad i = 1, \ldots, 2n, \quad \text{(by orthogonality).}
\end{aligned}$$

The latter equation and Eq. 4.84 give

$$\dot{q}_i(t) = \lambda_i\, q_i(t) + \frac{\left(\mathbf{v}^T\,\mathbf{b}\,\mathbf{f}(t)\right)_i}{\mathbf{v}_i^T\,\mathbf{u}_i}, \quad i = 1, \ldots, 2n, \tag{4.88}$$

whose solution is (see Appendix B)

$$q_i(t) = q_{i,0}\, e^{\lambda_i t} + \int_0^t e^{\lambda_i (t-s)}\, \frac{\left(\mathbf{v}^T\,\mathbf{b}\,\mathbf{f}(s)\right)_i}{\mathbf{v}_i^T\,\mathbf{u}_i}\, ds, \tag{4.89}$$

where $\{q_{i,0}\}$ are the initial values of $\{q_i(t)\}$, so that

$$\mathbf{z}(t) = \sum_{i=1}^{2n} \mathbf{u}_i\, q_i(t) = \sum_{i=1}^{2n} \mathbf{u}_i \left[q_{i,0}\, e^{\lambda_i t} + \int_0^t e^{\lambda_i (t-s)}\, \frac{\left(\mathbf{v}^T\,\mathbf{b}\,\mathbf{f}(s)\right)_i}{\mathbf{v}_i^T\,\mathbf{u}_i}\, ds \right]. \tag{4.90}$$

Note that the coefficients $\{\lambda_i\}$ and the forcing functions $\{\left(\mathbf{v}^T\,\mathbf{b}\,\mathbf{f}(t)\right)_i / \left(\mathbf{v}_i^T\,\mathbf{u}_i\right)$ of Eq. 4.88 are complex-valued and so are the solution $\{q_i(t)\}$. It can be shown by using properties of eigenvectors and eigenvalues that the state vector $\mathbf{z}(t)$ has real-valued components [1]. We will only mention that the components of this vector are real as they represent physical quantities, displacements, and velocities of the system masses.

If the real parts of the eigenvalues $\{\lambda_i\}$, $i = 1, \ldots, 2n$, are negative, the system solution becomes

$$\mathbf{z}_{SS}(t) \simeq \sum_{i=1}^{2n} \mathbf{u}_i \int_0^t e^{\lambda_i (t-s)} \frac{\left(\mathbf{v}^T \mathbf{b}\, \mathbf{f}(s)\right)_i}{\mathbf{v}_i^T \mathbf{u}_i} ds, \tag{4.91}$$

for large times, and is referred to as the ***steady-state*** solution.

The initial conditions for the modal coordinates $\{q_i(t)\}$ result from the initial conditions $(\mathbf{x}_0, \dot{\mathbf{x}}_0)$ in the physical space and the representation of $\mathbf{z}(t)$ in Eq. 4.87. We have $\mathbf{z}_0 = \mathbf{u}\,\mathbf{q}_0$ at time $t = 0$ from the initial conditions in the physical space. The left multiplication of $\mathbf{z}_0 = \mathbf{u}\,\mathbf{q}_0$ by $\mathbf{v}^T$ gives $\mathbf{v}^T\,\mathbf{z}_0 = \mathbf{v}^T\,\mathbf{u}\,\mathbf{q}_0 = \text{diag}\{\mathbf{v}_i^T\,\mathbf{u}_i\,q_{i,0}\}$ or $\left(\mathbf{v}^T\,\mathbf{z}_0\right)_i = \mathbf{v}_i^T\,\mathbf{u}_i\,q_{i,0}$. This shows that the initial conditions for the modal coordinates are

$$q_{i,0} = \frac{\left(\mathbf{v}^T\,\mathbf{z}_0\right)_i}{\left(\mathbf{v}_i^T\,\mathbf{u}_i\right)}, \qquad i = 1, \ldots, 2n. \tag{4.92}$$

4.5.5 Free Vibration

The free vibration solution results from Eq. 4.90 by setting $\mathbf{f}(t) = \mathbf{0}$. For completeness, we also present the solution by direct arguments for this special case. The representation of $\mathbf{z}(t)$ in Eq. 4.87 and the equation of motion $\dot{\mathbf{z}}(t) = \mathbf{a}\,\mathbf{z}(t)$ give

$$\begin{aligned}
&\mathbf{u}\,\dot{\mathbf{q}}(t) = \mathbf{a}\,\mathbf{u}\,\mathbf{q}(t) \quad (\text{ which becomes}) \\
&\mathbf{v}^T\,\mathbf{u}\,\dot{\mathbf{q}}(t) = \mathbf{v}^T\,\mathbf{a}\,\mathbf{u}\,\mathbf{q}(t) \quad (\text{by left multiplication with } \mathbf{v}^T) \\
&\mathbf{v}_i^T\,\mathbf{u}_i\,\dot{q}_i(t) = \mathbf{v}_i^T\,\mathbf{a}\,\mathbf{u}_i\,q_i(t), \quad i = 1, \ldots, 2n, \quad (\text{ by orthogonality}).
\end{aligned}$$

The latter condition gives

$$\begin{aligned}
&\dot{q}_i(t) = \lambda_i\,q_i(t) \quad \text{so that } q_i(t) = q_{i,0}\,e^{\lambda_i t} \\
&\mathbf{z}(t) = \sum_{i=1}^{2n} \mathbf{u}_i\,q_{i,0}\,e^{\lambda_i t},
\end{aligned} \tag{4.93}$$

where $\{q_{i,0}\}$ denotes initial conditions for $\{q_i(t)\}$, which can be obtained by the approach of the previous subsection.

Example 4.9 Consider a SDOF system with damping ratio $\zeta = 0.05$ and natural frequency $\omega = 6$ rad/s, which is subjected to a unit force applied suddenly. The oscillator is at rest at the initial time. Its displacement $x(t)$ satisfies the differential equation

$$\ddot{x} + 2\,\zeta\,\omega\,\dot{x} + \omega^2\,x = 1, \quad t \geq 0,$$

with zero initial conditions, and has the expression (see Example 2.5)

$$x(t) = \frac{1}{\omega^2}\left[e^{-\zeta\,\omega\,t}\left(-\frac{\zeta}{\sqrt{1-\zeta^2}}\,\sin(\omega_d\,t) - \cos(\omega_d\,t)\right) + 1\right].$$

We also calculate this solution by using the state-space representation of Eq. 4.77. The components of the state vector $\mathbf{z}(t)$ are the oscillator displacement $x(t)$ and velocity $\dot{x}(t)$. The matrices $\mathbf{a}$ and $\mathbf{b}$ and the forcing function are

$$\mathbf{a} = \begin{bmatrix} 0 & 1 \\ -\omega^2 & -2\,\zeta\,\omega \end{bmatrix}, \quad \mathbf{b} = \begin{bmatrix} 0 \\ 1 \end{bmatrix}$$

and $f(t) = 1$. Of course, there is no practical reason for these calculations. They are presented to illustrate the state-space approach in a simple setting. Its implementation involves the following three steps.

- *Step 1*: Find the eigenvalues and the right/left eigenvectors of the system matrix by using the MATLAB functions $[\mathbf{u}, \mathbf{d}] = \text{eig}(\mathbf{a})$ and $[\mathbf{v}, \mathbf{d}] = \text{eig}(\mathbf{a}')$. These system properties have been calculated in Example 4.8 for $\omega = 6$ and $\zeta = 0.05$.
- *Step 2*: Construct the differential equations for $\{q_i(t)\}$ (see Eq. 4.88), i.e., the equations $\dot{q}_i(t) = \lambda_i\,q_i(t) + d_i$, where $d_i = 0.5069 \pm 0.254\,i$, $i = 1, 2$, have been obtained from

$$d_i = \frac{\left(\mathbf{v}^T\,\mathbf{b}\,\mathbf{f}(t)\right)_i}{\mathbf{v}_i^T\,\mathbf{u}_i} = \frac{v_{i,2}}{\mathbf{v}_i^T\,\mathbf{u}_i}, \qquad i = 1, 2,$$

  where $v_{i,2}$ denotes the second component of $\mathbf{v}_i$. The solutions of these types of equations are discussed in Example B.2 (Appendix B). Since the forcing functions $\{d_i\}$ are constants, we can obtain these solutions by simple calculations. Recall that the general solution of these equations has the form

$$q_i(t) = c_i\,e^{\lambda_i\,t} + q_{p,i}(t),$$

  where $c_i\,e^{\lambda_i\,t}$ is the general solution of the homogeneous equation, c_i is a constant, and $q_{p,i}(t)$ denotes a particular solution of the inhomogeneous equation. Since d_i does not depend on time, $q_{p,i}(t) = -d_i/\lambda_i$, and $c_i = d_i/\lambda_i$ for the initial condition $\mathbf{z}(0) = \mathbf{0}$. The resulting expressions of the modal coordinates and system solution are

$$q_i(t) = \frac{d_i}{\lambda_i}\left(e^{\lambda_i\,t} - 1\right) \quad \text{and} \quad \mathbf{z}(t) = \sum_{i=1}^{2}\mathbf{u}_i\,\frac{d_i}{\lambda_i}\left(e^{\lambda_i\,t} - 1\right).$$

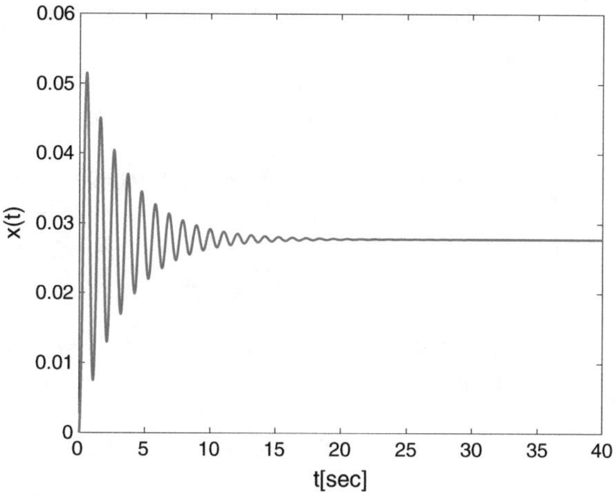

Fig. 4.13 Displacement of a SDOF system subjected to a suddenly applied unit force

The first component of $\mathbf{z}(t)$ is the oscillator displacement. This component and the expression of the oscillator displacement $x(t)$ given above are plotted in Fig. 4.13 for $\omega = 6$ and $\zeta = 0.05$. The two solutions are indistinguishable at the scale of the figure.

Example 4.10 A 2DOF system with mass, stiffness, and damping matrices

$$\mathbf{m} = \begin{bmatrix} 3000 & 0 \\ 0 & 2000 \end{bmatrix} \text{ kg}, \quad \mathbf{k} = \begin{bmatrix} 18 & -8 \\ -8 & 8 \end{bmatrix} \text{ kN/m and } \mathbf{c} = \begin{bmatrix} 3 & -1 \\ -1 & 1 \end{bmatrix} \text{ kN.sec/m}$$

is subjected to the force $f(t) = \sin(t)$ kN applied to the first degree of freedom. The system is at rest at the initial time. We apply the state-space method to find the displacement vector $\mathbf{x}(t)$ of the system whose components $x_1(t)$ and $x_2(t)$ are the displacements at the two degrees of freedom. The state vector $\mathbf{z}(t)$ in Eq. 4.78 is four-dimensional and satisfies the equation (see Eq. 4.77)

$$\dot{\mathbf{z}}(t) = \begin{bmatrix} \mathbf{0} & \mathbf{I} \\ -\mathbf{m}^{-1}\mathbf{k} & -\mathbf{m}^{-1}\mathbf{c} \end{bmatrix} \mathbf{z}(t) + \begin{bmatrix} \mathbf{0} \\ \mathbf{m}^{-1} \end{bmatrix} \begin{bmatrix} 1 \\ 0 \end{bmatrix} \sin(t),$$

where

$$\mathbf{a} = \begin{bmatrix} \mathbf{0} & \mathbf{I} \\ -\mathbf{m}^{-1}\mathbf{k} & -\mathbf{m}^{-1}\mathbf{c} \end{bmatrix}, \quad \mathbf{b} = \begin{bmatrix} \mathbf{0} \\ \mathbf{m}^{-1} \end{bmatrix} \text{ and } \mathbf{f}(t) = \begin{bmatrix} 1 \\ 0 \end{bmatrix} \sin(t).$$

The right and left eigenvectors of $\mathbf{a}$ in Eq. 4.86 are

$$\mathbf{u} = \begin{bmatrix} 0.7812+0.0000\,i & 0.7812+0.0000\,i & 0.7543+0.0000\,i & 0.7543+0.0000\,i \\ -0.5572-0.0097\,i & -0.5572+0.0097\,i & 0.3480-0.0730\,i & 0.3480+0.0730\,i \\ 0.0623+0.2215\,i & 0.0623-0.2215\,i & 0.1048+0.4891\,i & 0.1048-0.4891\,i \\ -0.0297-0.1591\,i & -0.0297+0.1591\,i & 0.0580+0.2260\,i & 0.0580-0.2260\,i \end{bmatrix}$$

$$\mathbf{v} = \begin{bmatrix} -0.0169-0.1190\,i & -0.0169+0.1190\,i & -0.0591-0.3156\,i & -0.0591+0.3156\,i \\ 0.0579+0.2468\,i & 0.0579-0.2468\,i & -0.0649-0.4447\,i & -0.0649+0.4447\,i \\ 0.4094+0.0368\,i & 0.4094-0.0368\,i & 0.4843-0.0195\,i & 0.4843+0.0195\,i \\ -0.8674+0.0000\,i & -0.8674+0.0000\,i & 0.6783+0.0000\,i & 0.6783+0.0000\,i. \end{bmatrix}$$

The modal coordinates $\{q_i(t)\}$ satisfy Eq. 4.88 with the forcing functions given by the components

$$\begin{bmatrix} 0.1629+0.0414\,i \\ 0.1629-0.0414\,i \\ 0.2083+0.0321\,i \\ 0.2083-0.0321\,i \end{bmatrix} \sin(t)\ \text{kN.s/m}$$

of the vector $\mathbf{v}^T\,\mathbf{b}\,\mathbf{f}(t)$ scaled by the non-zero entries of the diagonal matrix $\mathbf{v}^T\,\mathbf{u}$. Their solutions in Eq. 4.89 and the representation of $\mathbf{z}(t)$ in Eq. 4.90 deliver the displacement and velocity functions of the system masses. The solid and dotted lines of Fig. 4.14 show the displacements of the first and second degrees of freedom.

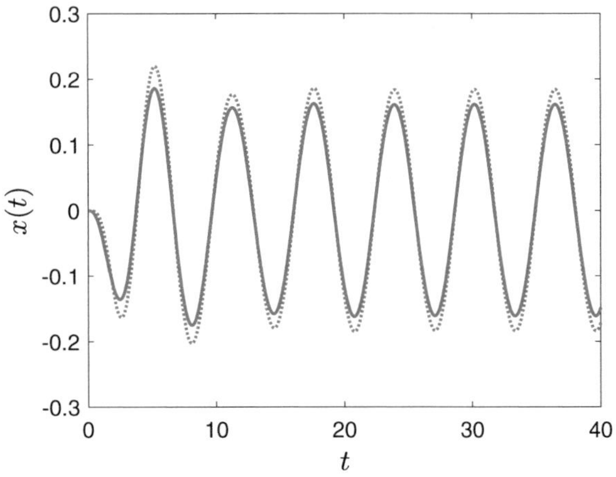

Fig. 4.14 Displacements of the first and second degrees of freedom (solid and dotted lines)

4.6 Frequency Domain Analysis

We extend developments for SDOF systems related to the Fourier series representations to MDOF systems with proportional and non-proportional damping. First, we construct Fourier series for vector-valued forcing functions.

4.6.1 *Fourier Series Representation of Vector-Valued Forcing Functions*

Suppose that the applied force $\mathbf{f}(t)$ is a periodic vector-valued function with period τ, i.e., $\mathbf{f}(t) = \mathbf{f}(t+\tau)$ for any t in the domain of definition of $\mathbf{f}(t)$. The components $\{f^{(r)}(t)\}$ of this n-dimensional function admit the representation (see Eq. 2.52)

$$f^{(r)}(t)=\frac{a_0^{(r)}}{2}+\sum_{j=1}^{\infty}\left(a_j^{(r)}\cos(\nu_j t)+b_j^{(r)}\sin(\nu_j t)\right),\quad r=1,\ldots,n, \tag{4.94}$$

where $\nu_1 = 2\pi/\tau$ and $\nu_j = j\,\nu_1$, $j = 0, 1, 2, \ldots$, are multiple of ν_1.

The coefficients of the above representation of the real-valued functions $\{f^{(r)}(t)\}$ result from Eq. 2.53 with $f^{(r)}(t)$ in place of $f(t)$. For calculations, the Fourier series representations of Eq. 4.94 are truncated by retaining the top m terms so that $f^{(r)}(t)$ is approximated by

$$f^{(r)}(t)\simeq\tilde{f}^{(r)}(t)=\frac{a_0^{(r)}}{2}+\sum_{j=1}^{m}\left(a_j^{(r)}\cos(\nu_j t)+b_j^{(r)}\sin(\nu_j t)\right),$$
$$r=1,\ldots,n, \tag{4.95}$$

and

$$\mathbf{f}(t)\simeq\tilde{\mathbf{f}}(t)=\frac{1}{2}\,\mathbf{a}_0+\sum_{j=1}^{m}\left(\mathbf{a}_j\cos(\nu_j t)+\mathbf{b}_j\sin(\nu_j t)\right), \tag{4.96}$$

where $\mathbf{a}_0$, $\mathbf{a}_j$, and $\mathbf{b}_j$, $j = 1, \ldots, m$, are n-dimensional column vectors with components $\{a_0^{(r)}\}$, $\{a_j^{(r)}\}$, and $\{b_j^{(r)}\}$. Note that, the components of the truncated version $\tilde{\mathbf{f}}(t)$ of the periodic forcing function $\mathbf{f}(t)$ have energy at the same frequencies, the discrete frequencies $\{\nu_j\}$, $j = 0, 1, \ldots, m$. Generally, the energy associated with these frequencies differs for different components of $\tilde{\mathbf{f}}(t)$. The Fourier transforms of the components of $\tilde{\mathbf{f}}(t)$ have the form in Eq. 2.57.

The following subsections find the response of MDOF systems with proportional and non-proportional damping subjected to vector-valued forcing functions represented by truncated Fourier series.

4.6.2 Steady-State Solution: Proportional Damping

Suppose that the vector-valued forcing function $\mathbf{f}(t)$ in Eq. 4.21 can be represented by the truncated Fourier series $\tilde{\mathbf{f}}(t)$ given by Eq. 4.96. The modal coordinates $\{q_i(t)\}$ are the solutions of Eq. 4.26 with $\tilde{f}_i(t) = \left(\boldsymbol{\Psi}^T \tilde{\mathbf{f}}(t)\right)_i$ in place of $f_i(t) = \left(\boldsymbol{\Psi}^T \mathbf{f}(t)\right)_i$, i.e.,

$$f_i(t) \simeq \tilde{f}_i(t) = \left(\boldsymbol{\Psi}^T \tilde{\mathbf{f}}(t)\right)_i = \frac{\alpha_{i,0}}{2} + \sum_{j=1}^{m} \left(\alpha_{i,j} \cos(\nu_j t) + \beta_{i,j} \sin(\nu_j t)\right), \tag{4.97}$$

where

$$\alpha_{i,0} = \left(\boldsymbol{\Psi}^T \mathbf{a}_0\right)_i, \quad \alpha_{i,j} = \left(\boldsymbol{\Psi}^T \mathbf{a}_j\right)_i, \quad \text{and } \beta_{i,j} = \left(\boldsymbol{\Psi}^T \mathbf{b}_j\right)_i,$$
$$i = 1, \ldots, n, \quad j = 1, \ldots, m. \tag{4.98}$$

The steady-state modal coordinates have the expressions (see Eq. 2.59)

$$\tilde{q}_{ss,i}(t) = \frac{\alpha_{i,0}}{2\, \tilde{m}_i\, \tilde{k}_i} + \sum_{j=1}^{m} \left[\frac{\alpha_{i,j}}{\tilde{m}_i} \frac{r_{d,i}(\nu_j)}{\omega_i^2} \cos(\nu_j t - \varphi_j) + \frac{\beta_{i,j}}{\tilde{m}_i} \frac{r_{d,i}(\nu_j)}{\omega_i^2} \sin(\nu_j t - \varphi_j) \right], \tag{4.99}$$

where

$$r_{d,i}(\nu) = \frac{1}{\sqrt{\left((1 - (\nu/\omega_i)^2\right)^2 + \left(2\, \zeta_i\, \nu/\omega_i\right)^2}} \tag{4.100}$$

denotes the dynamic amplification factor (DAF) for mode i and

$$\tan(\varphi_j) = \frac{2\, \zeta_i\, \nu_j/\omega_i}{1 - (\nu_j/\omega_i)^2} \tag{4.101}$$

is the phase angle associated with frequency ν_j (see Eqs. 2.44 and 2.45). The amplitude phase representation of the steady-state solution has the expression

$$\tilde{q}_{ss,i}(t) = \frac{\alpha_{i,0}}{2\, \tilde{m}_i\, \tilde{k}_i} + \sum_{j=1}^{m} \frac{r_{d,i}(\nu_j)}{\tilde{m}_i\, \omega_i^2} \sqrt{\alpha_{i,j}^2 + \beta_{i,j}^2}\, \sin(\nu_j t - \varphi_j + \theta_j), \tag{4.102}$$

by using the notations

$$\sin(\theta_j) = \frac{\alpha_{i,j}}{\sqrt{\alpha_{i,j}^2 + \beta_{i,j}^2}} \quad \text{and} \quad \cos(\theta_j) = \frac{\beta_{i,j}}{\sqrt{\alpha_{i,j}^2 + \beta_{i,j}^2}} \tag{4.103}$$

so that $\theta_j = \tan^{-1}(\alpha_{i,j}/\beta_{i,j})$. Note that the Fourier transform of the steady-state solutions,

$$\text{FT}[\tilde{q}\ \text{ss},i](\nu) = \frac{\alpha_{i,0}}{2\,\tilde{m}_i\,\tilde{k}_i}\,\delta(\nu) + \sum_{j=1}^{m} \frac{r_{d,i}(\nu_j)}{\tilde{m}_i\,\omega_i^2}\sqrt{\alpha_{i,j}^2 + \beta_{i,j}^2}\,\delta(\nu - \nu_j),$$

$$i = 1, \ldots, n, \tag{4.104}$$

and the input $\tilde{f}(t)$ have energy at the same frequencies. However, the energies associated with the input and output frequencies differ and so do the energies of different modal coordinates.

4.6.3 Steady-State Solution: Non-proportional Damping

Consider the truncated Fourier series representation of the forcing function $\mathbf{f}(t)$ given by Eqs. 4.95 and 4.96. The forcing functions of Eq. 4.88 are

$$\frac{\left(\mathbf{v}^T\,\mathbf{b}\,\tilde{\mathbf{f}}(t)\right)_i}{\mathbf{v}_i^T\,\mathbf{u}_i} = \alpha_{i,0} + \sum_{j=1}^{m}\left(\alpha_{i,j}\,\cos(\nu_j\,t) + \beta_{i,j}\,\sin(\nu_j\,t)\right), \quad i = 1, \ldots, 2n, \tag{4.105}$$

where

$$\alpha_{i,0} = \frac{\left(\mathbf{v}^T\,\mathbf{b}\,\mathbf{a}_0\right)_i}{2\,\mathbf{v}_i^T\,\mathbf{u}_i}, \quad \alpha_{i,j} = \frac{\left(\mathbf{v}^T\,\mathbf{b}\,\mathbf{a}_j\right)_i}{\mathbf{v}_i^T\,\mathbf{u}_i}, \quad \text{and } \beta_{i,j} = \frac{\left(\mathbf{v}^T\,\mathbf{b}\,\mathbf{b}_j\right)_i}{\mathbf{v}_i^T\,\mathbf{u}_i},$$

$$i = 1, \ldots, 2n. \tag{4.106}$$

If the real parts of the eigenvalues $\{\lambda_i\}$ are negative, the modal coordinates reach the steady-state solution as time increases indefinitely so that for large times (see Eq. 4.89)

$$q_i(t) \simeq q_{\text{ss},i}(t) = \int_0^t e^{\lambda_i\,(t-s)}\,\frac{\left(\mathbf{v}^T\,\mathbf{b}\,\mathbf{f}(s)\right)_i}{\mathbf{v}_i^T\,\mathbf{u}_i}\,ds, \quad i = 1, \ldots, 2n. \tag{4.107}$$

The steady-state system solution $\mathbf{z}(t)$ in Eq. 4.91 with $\mathbf{f}(t)$ approximated by $\tilde{\mathbf{f}}(t)$ has the form

$$\begin{aligned}
\tilde{\mathbf{z}}_{SS}(t) &\simeq \sum_{i=1}^{2n} \mathbf{u}_i \int_0^t e^{\lambda_i (t-s)} \left[\alpha_{i,0} + \sum_{j=1}^{m} \left(\alpha_{i,j} \cos(\nu_j s) + \beta_{i,j} \sin(\nu_j s) \right) \right] ds \\
&= \sum_{i=1}^{2n} \mathbf{u}_i \left[\alpha_{i,0} \int_0^t e^{\lambda_i (t-s)} ds \right. \\
&\quad \left. + \sum_{j=1}^{m} \left(\alpha_{i,j} \int_0^t e^{\lambda_i (t-s)} \cos(\nu_j s)\, ds + \beta_{i,j} \int_0^t e^{\lambda_i (t-s)} \sin(\nu_j s)\, ds \right) \right].
\end{aligned} \tag{4.108}$$

The steady-state solution $\tilde{\mathbf{z}}_{SS}(t)$ is available analytically via the definite integrals

$$\begin{aligned}
&\int e^{a\,t}\, dt = e^{a\,t}/a, \\
&\int e^{a\,t} \sin(b\,t)\, dt = e^{a\,t} \left(a \sin(b\,t) - b \cos(b\,t) \right)/(a^2 + b^2) \\
&\int e^{a\,t} \cos(b\,t)\, dt = e^{a\,t} \left(a \cos(b\,t) + b \sin(b\,t) \right)/(a^2 + b^2),
\end{aligned}$$

which hold for real and complex parameter a of the exponential function.

4.6.4 Matrix Exponential

The solution of the state-space equation of motion in Eq. 4.77 can be given in the form

$$\mathbf{z}(t) = \mathbf{\Gamma}(t)\, \mathbf{z}_0 + \int_0^t \Gamma(t-s)\, \mathbf{b}\, \mathbf{f}(s)\, ds, \quad t \geq 0, \tag{4.109}$$

where the $(2n, 2n)$-matrix $\mathbf{\Gamma}(t)$, referred to as ***transition matrix***, is the solution of the homogeneous version of Eq. 4.77, i.e., the differential equation

$$\dot{\mathbf{\Gamma}}(t) = \mathbf{a}\, \mathbf{\Gamma}(t), \quad t \geq 0, \tag{4.110}$$

with the initial condition $\mathbf{\Gamma}(0) = \mathbf{I}$ [1, Chap. 1] The first and second terms in the expression of $\mathbf{z}(t)$ given by Eq. 4.109 correspond to free and forced vibrations. The Duhamel's integral of Eq. 2.31 developed for SDOF systems is a special case of Eq. 4.109.

For time-invariant matrices $\mathbf{a}$, as considered in our study, the transition matrix can be represented by the power series

$$\boldsymbol{\Gamma}(t) = \mathbf{I} + \mathbf{a}\,t + \mathbf{a}^2\,t^2/2! + \cdots, \tag{4.111}$$

which converges uniformly and absolutely on bounded time intervals. The limit of this series is denoted by $\exp\big(\mathbf{a}\,t\big)$ and referred to as ***matrix exponential***. With this representation of the transition matrix, the solution $\mathbf{z}(t)$ in Eq. 4.109 takes the form

$$\mathbf{z}(t) = e^{\mathbf{a}\,t}\,\mathbf{z}_0 + \int_0^t e^{\mathbf{a}\,(t-s)}\,\mathbf{b}\,\mathbf{f}(s)\,ds, \quad t \geq 0. \tag{4.112}$$

The matrix exponential $\exp\big(\mathbf{a}\,t\big)$ can be calculated from truncated versions of its series representation given by Eq. 4.111. However, it is more convenient to use the MATLAB function $\mathrm{expm}\big(\mathbf{a}\,t\big)$, which gives this function. The output of this MATLAB function can be used to calculate the free and forced vibration terms in the expression of $\mathbf{z}(t)$.

Example 4.11 Consider the differential equation $\ddot{x}(t) = \sin(\nu\,t)$, $t \geq 0$, with the initial conditions $x(0) = x_0$ and $\dot{x}(0) = \dot{x}_0$. The integration of this equation gives $\dot{x}(t) = -\cos(\nu\,t)/\nu + A$ and $x(t) = -\sin(\nu\,t)/\nu^2 + A\,t + B$ so that $x_0 = B$, $\dot{x}_0 = -1/\nu + A$ and

$$x(t) = x_0 + \big(\dot{x}_0 + 1/\nu\big)\,t - \sin(\nu\,t)/\nu^2, \quad t \geq 0.$$

The state-space representation of the differential equation $\ddot{x}(t) = \sin(\nu\,t)$, $t \geq 0$, has the form in Eq. 4.77 with

$$\mathbf{a} = \begin{bmatrix} 0 & 1 \\ 0 & 0 \end{bmatrix}, \quad \mathbf{b} = \begin{bmatrix} 0 \\ 1 \end{bmatrix}, \quad \text{and} \quad f(t) = \sin(\nu\,t).$$

The MATLAB solution of the first term of Eq. 4.112 is $\mathrm{expm}\big(\mathbf{a}\,t\big) * \mathbf{z}_0$, where $\mathbf{z}_0$ is a 2-dimensional vector with components x_0 and $\dot{x}_0$. The MATLAB solution of the integrand of the second term of Eq. 4.112 is $\mathrm{expm}\big(\mathbf{a}\,(t-s)\big) * \mathbf{b} * \sin(\nu\,t)$. These functions are integrated in the following MATLAB function.

```
function matrix_exp(x0,xd0,nu,tf,nt)
%
%   Solution of x''(t)=sin(nu*t) with ICs (x0,xd0)
%=================================================================
%   Direct integration
%-------------------------------------
dt=tf/nt; time=0:dt:tf; xt=x0+xd0*time+time/nu-sin(nu*time)
                                    /nu^2;
xdt=xd0+(1-cos(nu*time))/nu;
%==================================================
%   Matrix exponential
%------------------------------------------
a=[0 1;0 0]; a_exp=zeros(2,2,nt+1); for kt=1:nt+1
    at=[0 time(kt);0 0];
    a_exp(:,:,kt)=expm(at);
```

```
    x_hom(:,kt)=a_exp(:,:,kt)*[x0;xd0];
end
%----------------------------------------
%   Integrand for inhomogeneous solution
%----------------------------------------
x_inhom=zeros(2,nt+1); for kt=1:nt
    %------------------------------------
    %   Integral to a t in [0,tf]
    %----------------------------------
    sol=zeros(2,1);
    for ks=1:kt+1
        at_int=[0 time(kt+1)-time(ks);0 0];
        a_exp_int=expm(at_int);
        sol=sol+a_exp_int*[0;sin(nu*time(ks))]*dt;
    end
    x_inhom(:,kt+1)=sol;
end x_inhom=x_inhom+x_hom;
%----------------------
figure plot(time,x_hom)
xlabel('$t$','interpreter','latex','fontsize',18)
ylabel('$x_{\mathrm{hom}}(t)$','interpreter','latex',
       'fontsize',18) hold
plot(time,x0+xd0*time,':',time,xd0,':')
%-------------------
figure plot(time,x_inhom,'--', 'linewidth',2)
xlabel('$t$','interpreter','latex','fontsize',18)
ylabel('$x(t)$','interpreter','latex','fontsize',18) hold
plot(time,xt,'linewidth',2) plot(time,xdt,'linewidth',2)
%====================================================
%   matrix_exp(1,-1,6,5,50);
%
```

The heavy solid and dashed lines in Fig. 4.15 are the solutions $x(t)$ and $\dot{x}(t)$ by direct integration and by matrix exponential for $x_0 = 1$, $\dot{x}_0 = -1$, frequency $\nu = 6$, and time step $\Delta t = 0.1$. The two lines are nearly indistinguishable at the scale of the figure. We note that the solution of $\ddot{x}(t) = \sin(\nu\, t)$ by the state-space approach of the previous section is unstable since the matrix **a** associated with this equation has zero eigenvalues. In contrast, the solution by the matrix exponential does not encounter this difficulty. However, it is computationally less efficient than the state-space representation when dealing with large dynamical systems.

Example 4.12 Consider a damped single degree of freedom in free vibration with unit mass, natural frequency ω, and damping ratio ζ. The unit impulse response function of the system in Eq. 2.30 is the oscillator displacement function for the initial conditions $x_0 = 0$ and $\dot{x}_0 = 1$. It has the expression $h(t) = \exp(-\omega\, t)\, \sin(\omega_d\, t)/\omega_d$, where $\omega_d = \omega\sqrt{1-\zeta^2}$.

The free vibration solution of this SDOF system can also be obtained from Eq. 4.112 with $\mathbf{f}(t) = \mathbf{0}$ and $\mathbf{z}_0$ with components $x_0 = 0$ and $\dot{x}_0 = 1$. Accordingly, the oscillator displacement and velocity are $e_{12}(t)\,\dot{x}_0$, and $e_{22}(t)\,\dot{x}_0$, where $\mathbf{e}(t) = \{e_{ij}(t)\} = \exp(\mathbf{a}\, t)$. The solid and dashed lines of Fig. 4.16 show the unit impulse

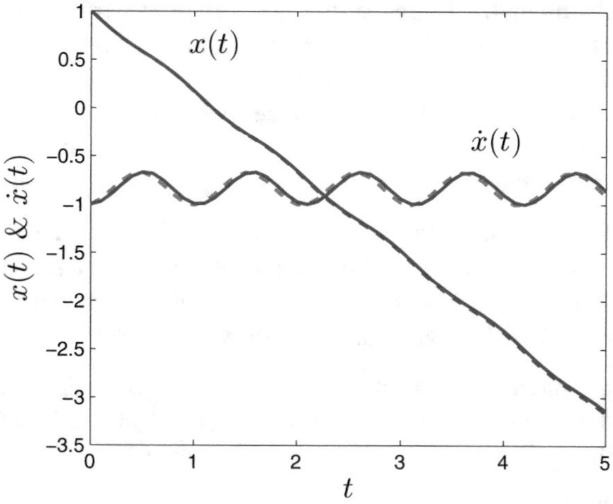

Fig. 4.15 Solution $x(t)$ and $\dot{x}(t)$ of $\ddot{x}(t) = \sin(\nu\, t)$ by direct integration and matrix exponential (solid and dashed lines)

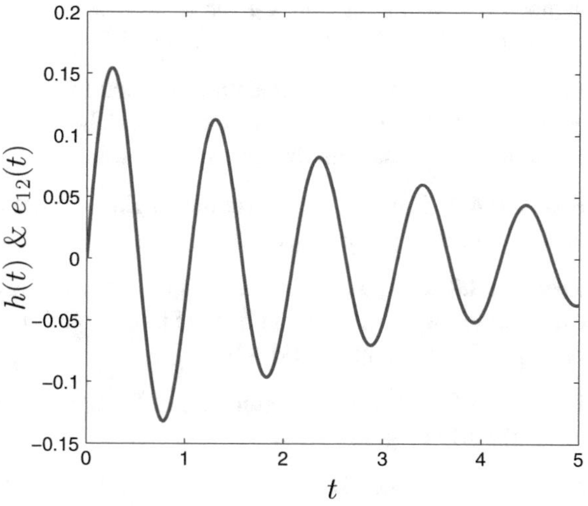

Fig. 4.16 Unit impulse response functions by physical considerations (Eq. 2.30) and matrix exponential (Eq. 4.112) (solid and dashed lines)

response function $h(t)$ in Eq. 2.30 and the displacement function $e_{12}(t)\,\dot{x}_0$ for $\omega = 6$, $\zeta = 0.05$, and $\dot{x}_0 = 1$. As expected, the two lines are indistinguishable at the scale of the figure.

4.7 Problems

Problem 4.1 Consider an undamped MDOF system in free vibration under the initial conditions $\mathbf{x}_0 = \mathbf{\Phi}_i$ and $\dot{\mathbf{x}}_0 = \mathbf{0}$, where $\mathbf{\Phi}_i$ denotes the ith mode of vibration. Show that the system vibrates in mode i.

Problem 4.2 Suppose that the 2-DOF system in Example 4.4 is subjected to an acceleration $a(t) = \sin(\nu t)$ applied at its fix end rather than the forcing function $f(t)$. First, find the steady-state modal and system responses for forcing frequencies ν in the range $[0, 1.5\,\omega_2]$. Then, pretend that the modal frequencies are unknown. Use the steady-state solutions to design an experiment for estimating the modal frequencies.

Problem 4.3 Plot the time evolution of the displacement vector $\mathbf{x}(t)$ of the 2-DOF system in Example 4.5 for a time interval $[0, \tau]$ and initial conditions of your choice. Plot also the modal shapes $\{\mathbf{\Phi}_i\}$ of this system and identify the projections $\{q_i(t)\}$ of $\mathbf{x}(t)$ on $\{\mathbf{\Phi}_i\}$ at an arbitrary time t.

Problem 4.4 Repeat the calculations of Example 4.3 and assume that the system is damped with damping matrix given by Eq. 4.14, $\alpha = 0.03$ and $\beta = 0.05$. Compare the system response with that in Example 4.3.

Problem 4.5 Repeat the calculations of Example 4.5 and assume that the system is damped with damping matrix given by Eq. 4.14, $\alpha = 0.03$ and $\beta = 0.05$. Compare the system response with that in Example 4.5.

Problem 4.6 Repeat the calculations of Example 4.4 and assume that the system is damped with damping matrix given by Eq. 4.14, $\alpha = 0.03$ and $\beta = 0.05$. Compare the system response with that in Example 4.4.

Problem 4.7 Show that the formulas of Eq. 4.85 are correct by using the orthogonality conditions given by Eq. 4.83.

Problem 4.8 Find the right and left eigenvectors of $\mathbf{a}$ in Example 4.8 by direct calculations and compare with MATLAB solutions. Show that the vectors are orthogonal in the sense of Eq. 4.83.

Problem 4.9 Show that solution $\mathbf{z}(t)$ of Sect. 4.5.4 does not depend on the scaling of the right/left eigenvectors.

Problem 4.10 Find the response of the 2-DOf system in Example 4.10 by the matrix exponential method.

References

1. R.W. Brockett, *Finite Dimensional Linear Systems* (John Wiley & Sons Inc., New York, 1970)
2. A.S. Cakmak, J.F Botha, W.G. Gray, *Computational and Applied Mathematics for Engineering Analysis* (Springer, New York, 1987)
3. T.K. Caughey, M.E.J. O'Kelly, Classical normal modes in damped linear dynamic systems. J. Appl. Mech. **32**, 583–588 (1965)
4. M.D. Greenberg, *Foundations of Applied Mathematics* (Prentice Hall Inc., Englewood Cliffs, 1978)

Chapter 5
Continuous Systems

It can be argued correctly that all mechanical/structural systems are continuous since there are no massless system components. Accordingly, they have to be viewed and analyzed as systems with infinite numbers of degrees of freedom. This observation may suggest that our work on SDOF/MDOF dynamical systems cannot be used to analyze continuous systems. This is not the case. There are two methods for analyzing continuous systems, and both methods involve approximations.

- *Method 1*: It approximates the solution of continuous systems by that of MDOF systems obtained by concentrating their distributed mass at finite numbers of points. There are no formulas delivering MDOF-based representations of continuous systems. Intuition and experience guide the construction of MDOF-based approximations. The accuracy of the method depends on the quality of the MDOF representation. The use of more familiar concepts is an advantage of the method. The number of degrees of freedom, which can be excessive for some systems, e.g., industrial pipes, and the difficulty to assess solution accuracy are its main disadvantages.
- *Method 2*: It uses equations of motions for continuous systems so that its implementation involves less simple mathematical tools. However, the method is conceptually similar to Method 1, i.e., solutions are represented by elements of linear spaces spanned by modal shapes, which are infinite sets of functions rather than finite sets of vectors. Theoretically, the method delivers the exact solution. Practically, we can only obtain approximations since the representation of system displacement has to be truncated, i.e., it has to be based on finite numbers of modal shapes for numerical solutions.

This section develops Method 2 for flexural and shear beams, see Figs. 5.1 and 5.3. The displacements $v(x, t)$ of these beams are real-valued functions of space and time arguments which satisfy partial differential equations. The method

M. D. Grigoriu, *Linear Dynamical Systems*,
https://doi.org/10.1007/978-3-030-64552-6_5

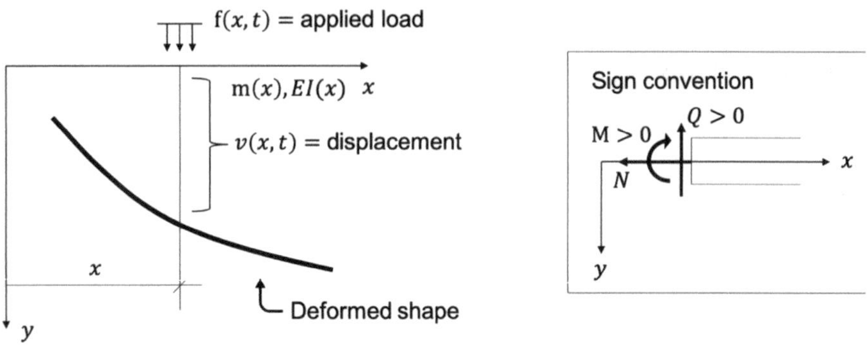

Fig. 5.1 Flexural beam and sign convention

of separation of variables is used for solution. This method views $v(x,t)$ as the product of two functions, $\varphi(x)$ and $q(t)$, which depend on only x and only t and satisfy ordinary differential equations. The non-trivial solutions of the homogeneous differential equation $\mathcal{D}[\varphi(x)] = 0$ for $\varphi(x)$, i.e., the eigenfunctions of the differential operator $\mathcal{D}$, are used to represent $v(x,t)$. The representation is similar to that for the solution of MDOF systems, which is based on eigenvectors (modal shapes).

For simplicity, we assume that the stiffness and the mass of the beams are space-invariant and that the systems are undamped. Also, the presentation is limited to time domain analysis. Our considerations on the frequency domain analysis for MDOF systems extend directly to the continuous systems considered in this chapter.

5.1 Flexural Beams

Consider the beam in Fig. 5.1 with mass per unit length $m(x)$ and stiffness $EI(x)$ under the action of a spatially distributed, time-dependent force $f(x,t)$. Denote the beam displacement at location x and time t by $v(x,t)$. The figure also shows the positive sign convention for the shear force $Q(x,t)$ and bending moment $M(x,t)$.

5.1.1 Physical System and Equations of Motion

This section lists without proof the differential relationships for flexural beams. Their derivation can be found in, e.g., [3]. These relationships also give the equations of motion for this continuous system.

$$\frac{\partial Q}{\partial x} = -\text{“}f\text{''}, \quad \text{where “}f\text{''} = f - m(x)\,\frac{\partial^2 v}{\partial t^2}$$

$$\frac{\partial M}{\partial x} = Q$$

$$\frac{\partial^2 v}{\partial x^2} = -\frac{M}{E\,I}$$

$$\frac{\partial^4 v}{\partial x^4} = -\frac{\partial^2}{\partial x^2}\left(\frac{M}{E\,I}\right) \tag{5.1}$$

If the stiffness $E\,I$ is constant, we have

$$\frac{\partial^4 v}{\partial x^4} = \frac{\text{“}f\text{''}}{E\,I} = \frac{f}{E\,I} - \frac{m(x)}{E\,I}\,\frac{\partial^2 v}{\partial t^2}. \tag{5.2}$$

The latter equation defines the beam displacement $v(x,t)$. Note that the applied force f in the first equality of Eq. 5.1 is augmented with the inertia force $-m(x)\,\partial^2 v/\partial t^2$. This is not a new concept. Recall the equation of motion for an undamped SDOF system, i.e., $m\,\ddot{x} + k\,x = f$. This equation can be written as $k\,x = f - m\,\ddot{x} = \text{“}f\text{''}$, which has the form of $\partial Q/\partial x = -\text{“}f\text{''}$.

5.1.2 Modal Shapes and Frequencies

The beam equation for the free vibration problem ($f = 0$) is

$$\frac{\partial^4 v}{\partial x^4} = -\frac{m}{E\,I}\,\frac{\partial^2 v}{\partial t^2} \qquad \text{or, equivalently,}\ \ v^{IV} + \frac{m}{E\,I}\,\ddot{v} = 0, \tag{5.3}$$

where the primes and dots denote partial derivatives with respect to the spatial and temporal arguments x and t. The method of separation of variables, which is the standard method for solving partial differential equations [1] (Chap. 5), represents the solution by

$$v(x,t) = \varphi(x)\,q(t), \tag{5.4}$$

where $\varphi(x)$ and $q(t)$ are functions of only space and time. This representation and Eq. 5.3 give

$$\varphi^{IV}\,q + \frac{m}{E\,I}\,\varphi\,\ddot{q} = 0 \quad \text{or, equivalently,}\ \ \frac{E\,I}{m}\,\frac{\varphi^{IV}}{\varphi} = -\frac{\ddot{q}}{q} = \omega^2 > 0. \tag{5.5}$$

We proceed by using the following standard arguments when dealing with partial differential equations, to clarify the notation ω^2, a positive constant.

- The terms $(E\,I/m)\,(\varphi^{IV}/\varphi)$ and $\ddot{q}/q$ are functions of only x and t, respectively. They must be constant for the following reason. Suppose we fix x so that $(E\,I/m)\,(\varphi^{IV}/\varphi)$ is a constant. Then, $\ddot{q}/q$ must take the same value at all times, which is equal to that of $(E\,I/m)\,(\varphi^{IV}/\varphi)$ for the selected x. A similar argument shows that $(E\,I/m)\,(\varphi^{IV}/\varphi)$ does not change with x for an arbitrary fixed time t. We denote the constant value of these terms by ω^2.
- The constant value of $(E\,I/m)\,(\varphi^{IV}/\varphi)$ and $-\ddot{q}/q$ must be strictly positive for the following reason. Consider the differential equation of $q(t)$ given by Eq. 5.5. If the constant is strictly positive, then $q(t)$ satisfies the equation $\ddot{q} + \omega^2\,q = 0$ whose solution is $q(t) = A\,\cos(\omega\,t) + B\,\sin(\omega\,t)$. If we set the constant in Eq. 5.5 to be strictly negative, i.e., $-\omega^2$ in place of ω^2, then $q(t)$ satisfies $\ddot{q} - \omega^2\,q = 0$ whose solution is $q(t) = C_1\,\exp(\omega\,t) + C_2\,\exp(-\omega\,t)$. The latter solution is physically unrealizable since it converges to $\pm\infty$ as $t \to \infty$, although there is no input (free vibration). In contrast, the solution corresponding to the positive constant ω^2 is oscillatory in agreement with physics.

The equalities of Eq. 5.5 give the following two ordinary differential equations:

$$\ddot{q}(t) + \omega^2\,q(t) = 0 \quad \text{(Initial value problem)} \quad \text{and}$$
$$\mathcal{D}[\varphi(x)] = \varphi^{IV}(x) - \frac{m\,\omega^2}{E\,I}\,\varphi(x) = 0 \quad \text{(Boundary value problem)} \tag{5.6}$$

for the components of $\varphi(x)$ and $q(t)$ of the displacement function $v(x,t)$. The solutions of these equations are

$$q(t) = A\,\cos(\omega\,t) + B\,\sin(\omega\,t) \quad \text{and}$$
$$\varphi(x) = C_1\,\sin(\beta\,x) + C_2\,\cos(\beta\,x) + C_3\,\sinh(\beta\,x) + C_4\,\cosh(\beta\,x), \tag{5.7}$$

where $\sinh(\alpha) = \left(e^{\alpha} - e^{-\alpha}\right)/2$, $\cosh(\alpha) = \left(e^{\alpha} + e^{-\alpha}\right)/2$, and

$$\beta^4 = \frac{m\,\omega^2}{E\,I}. \tag{5.8}$$

Definition 3 The non-trivial solutions of the (homogeneous) differential equation of $\varphi(x)$ are the ***eigenfunctions*** of the differential operator $\mathcal{D} = d^4/d\,x^4 - \beta^4$ of this equation. The values of ω^2, which determine the values of β in Eq. 5.8, for which non-trivial solutions exist are called the ***eigenvalues*** of $\mathcal{D}$. In our context, the eigenfunctions and eigenvalues of $\mathcal{D}$ have the physical meaning of ***modal shapes*** and ***modal frequencies*** for the continuous system under consideration.

The constants C_1, C_2, C_3, and C_4 result from the boundary conditions. The values of the parameter ω are such that the corresponding solutions $\varphi(x)$ are not trivial.

The initial conditions and the selected values of ω are used to find the constants A and B. The following example illustrates the construction of the modal shapes and frequency for a simple continuous system.

Example 5.1 Consider a simply supported beam with constant stiffness $E\,I$, mass per unit length m, and span l. The two ends of the beam are at $x = 0$ and $x = l$. Accordingly, the boundary conditions are $v(0,t) = v(l,t) = 0$ and $v''(0,t) = v''(l,t) = 0$ as the displacements and the bending moments (see the relationship between $\partial^2 v/\partial x^2$ and M given by Eq. 5.1) must be zero at the beam ends. Since boundary conditions must be satisfied at all times, we have (see the expression of $\varphi(x)$ in Eq. 5.7)

$$\text{Boundary condition at } x = 0:$$

$$v(0,t) = \varphi(0)\,q(t) = 0 \Longrightarrow \varphi(0) = 0 \Longrightarrow C_2 + C_4 = 0$$

$$\frac{\partial^2 v}{\partial x^2}(0,t) = \varphi''(0)\,q(t) = 0 \Longrightarrow \varphi''(0) = 0 \Longrightarrow -\beta^2\,C_2 + \beta^2\,C_4 = 0,$$

$$\text{so that } C_2 = C_4 = 0 \text{ and } \varphi(x) = C_1\,\sin(\beta\,x) + C_3\,\sinh(\beta\,x)$$

$$\text{Boundary condition at } x = l:$$

$$v(l,t) = \varphi(l)\,q(t) = 0 \Longrightarrow \varphi(l) = 0 \Longrightarrow C_1\,\sin(\beta\,l) + C_3\,\sinh(\beta\,l) = 0$$

$$\frac{\partial^2 v}{\partial x^2}(l,t) = \varphi''(l)\,q(t) = 0 \Longrightarrow \varphi''(l) = 0 \Longrightarrow -C_1\,\beta^2\,\sin(\beta\,l) + C_3\,\beta^2\,\sinh(\beta\,l) = 0.$$

The sum $2\,C_3\,\beta^2\,\sinh(\beta\,l) = 0$ of the last two equations implies $C_3 = 0$ since $\sinh(\beta\,l) \neq 0$, so that we have $C_2 = C_3 = C_4 = 0$ and $C_1\,\sin(\beta\,l) = 0$. Since C_1 cannot be zero for non-zero initial conditions (otherwise $v(x,t) = 0$ at all times), we require $\sin(\beta\,l) = 0$. This equation has the infinite number of solutions $\beta_n\,l = n\,\pi$, $n = 1, 2, \ldots$, which give (see the definition of β in Eq. 5.8)

$$(n\,\pi)^4 = (\beta_n\,l)^4 = \frac{m\,\omega_n^2\,l^4}{E\,I}, \qquad n = 1, 2, \ldots,$$

so that Eq. 5.5 has non-trivial solutions for

$$\omega_n = \left(\frac{n\,\pi}{l}\right)^2 \sqrt{\frac{E\,I}{m}}, \qquad n = 1, 2, \ldots. \tag{5.9}$$

The corresponding solutions $\varphi(x)$ are ($C_1 \neq 0$ and $C_2 = C_3 = C_4 = 0$)

$$\varphi_n(x) = \sin\left(\beta_n\,x\right) = \sin\left(\frac{n\,\pi}{l}\,x\right), \qquad n = 1, 2, \ldots, \tag{5.10}$$

and can be determined up to a multiplicative constant as $C_1 \neq 0$ remains undetermined. This constant is not written in Eq. 5.10.

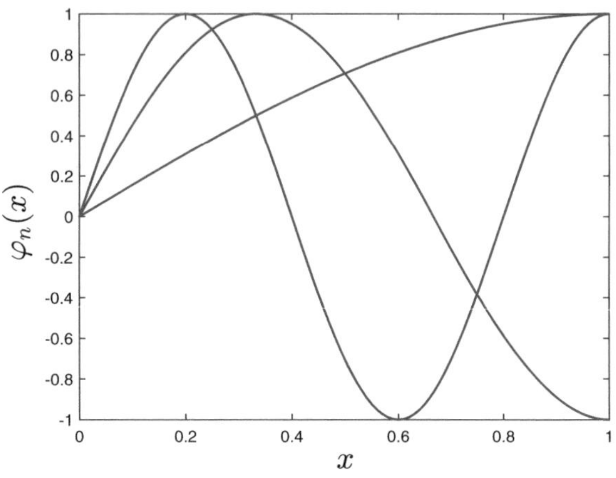

Fig. 5.2 First three modal shapes of a simply supported beam with $l = 1$

We note that the same result can be obtained by considering the latter two boundary conditions simultaneously, which give the homogeneous system of equations

$$\begin{bmatrix} \sin(\beta\, l) & \sinh(\beta\, l) \\ -\sin(\beta\, l) & \sinh(\beta\, l) \end{bmatrix} \begin{bmatrix} C_1 \\ C_3 \end{bmatrix} = \mathbf{0}. \tag{5.11}$$

Since the trivial solution $C_1 = C_3 = 0$ is not acceptable, we set zero the determinant of the above system of equations to identify the set of non-trivial solutions. This gives $2\,\sin(\beta\, l)\,\sinh(\beta\, l) = 0$ or $\sin(\beta\, l) = 0$ since $\sinh(\beta\, l) \neq 0$, as previously.

According to our definition, the (countable) infinite set of values $\{\omega_n\}$ of ω given by Eq. 5.9 for which the differential equation of $\varphi(x)$ admits non-trivial solutions are the beam ***modal frequencies***. The solutions $\{\varphi_n(x)\}$ in Eq. 5.10 corresponding to $\{\omega_n\}$ are the beam ***modal shapes***. Note that, as for MDOF systems, the modal shapes and frequencies are system properties. They are completely determined by the system mechanical properties, topology, and boundary conditions.

The first three modal shapes given by Eq. 5.10 are shown in Fig. 5.2. As expected, the modes satisfy the boundary conditions, i.e., $\varphi_n(0) = \varphi_n(l) = 0$ and $\varphi_n''(0) = \varphi_n''(l) = 0$, although the second set of conditions is less obvious from the plot.

We conclude with the following comments on properties of the modal shapes and the solutions of continuous systems.

1. **The modal shapes are orthogonal** in the sense that

$$\int_0^l \varphi_r(x)\,\varphi_n(x)\,dx = \int_0^l \sin(r\,\pi\,x/l)\,\sin(n\,\pi\,x/l)\,dx = \frac{l}{2}\,\delta_{rn}. \tag{5.12}$$

Note that this condition constitutes the continuous version of the orthogonality property of vectors in $\mathbb{R}^n$, which states two vectors are orthogonal if the sum of the products of their components is zero (see Appendix C). This condition is similar to that of Eq. 5.12, which adds the components of the "vectors" $\varphi_r(x)$ and $\varphi_n(x)$, i.e., the values of these functions at $x \in [0, l]$. The beam mass and stiffness are not present in Eq. 5.12 since they do not depend on x.

2. **The modal responses**,

$$q_n(t)\,\varphi_n(x) = \big(A_n\,\cos(\omega_n\,t) + B_n\,\sin(\omega_n\,t)\big)\,\sin\big(n\,\pi\,x/l\big), \quad n = 1, 2, \ldots, \tag{5.13}$$

are solutions of Eq. 5.3 for any constants A_n and B_n, so that

$$\tilde{v}(x,t) = \sum_{n=1}^{\infty} \varphi_n(x)\,q_n(t), \quad 0 \le x \le l, \quad t \ge 0, \tag{5.14}$$

also satisfies Eq. 5.3 since the equation of motion is linear. We use temporarily the notation $\tilde{v}(x,t)$ since it is not obvious that the beam displacement $v(x,t)$ admits this representation.

3. **The beam displacement** $v(x,t)$ admits the representation given by Eq. 5.14, i.e., the beam displacement $v(x,t)$ can be represented by the following sum

$$v(x,t) = \sum_{n=1}^{\infty} \varphi_n(x)\,q_n(t), \quad 0 \le x \le l, \quad t \ge 0, \tag{5.15}$$

of modal shapes weighted by the time-dependent coefficients $\{q_n(t)\}$. We only present an intuitive argument in support of this statement. It is based on the observation that the beam displacement $v(x,t)$ is an element of the linear space spanned by the modal shapes $\{\varphi_n(x)\}$. Suppose that the mass of a continuous system is concentrated at a refining set of points $\mathcal{P}_\alpha$, $\alpha = 1, 2, \ldots$, which means that the points of $\mathcal{P}_\alpha$ are included in the larger set of points $\mathcal{P}_{\alpha+1}$. The solutions of the MDOF systems defined by $\mathcal{P}_\alpha$ provide approximations of increasing accuracy for $v(x,t)$ as α increases. These approximations can be represented by weighted sums of modal shapes (see Eq. 4.19). It is expected that the beam displacement $v(x,t)$ admits the representation in Eq. 5.15 since it can be approximated to any accuracy by the displacements of MDOF systems for sufficiently large numbers of masses. Rigorous arguments on the validity of the modal representation of $v(x,t)$ involve less familiar concepts that are beyond the scope of this book. Interested reader can consult [2, Sect. III.6].

4. **The coefficients** $\{q_n(t)\}$ are the projections of $v(x,t)$ on the modal shapes, which can be calculated from

$$\int_0^l v(x,t)\,\varphi_m(x)\,dx = \sum_{n=1}^{\infty} q_n(t) \int_0^l \sin(m\,\pi\,x/l)\,\sin(n\,\pi\,x/l)\,dx = \frac{l}{2}\,q_m(t). \tag{5.16}$$

It can be shown that the term-by-term integration used to obtain the above equality is valid [2, Sect. II.11].

5.1.3 Forced Vibration

We have seen that the solution $v(x, t)$ can be viewed as an element of an infinite-dimensional linear space spanned by the eigenfunctions $\{\varphi_n(x)\}$ of the differential operator $\mathcal{D}[\varphi(x)] = 0$ in Eq. 5.6. We construct differential equations for the projections $\{q_n(t)\}$ of $v(x, t)$ on the basis functions $\{\varphi_n(x)\}$ by requiring that the displacement function $v(x, t)$ given by Eq. 5.15 satisfies the equation of motion $E\,I\,v^{IV} + m\,\ddot{v} = f$.

For the simply supported beam of Example 5.1, the equation of motion with $v(x, t)$ in Eq. 5.15 has the form

$$E\,I \sum_{n=1}^{\infty} q_n(t)\,(n\,\pi/l)^4\,\sin(n\,\pi\,x/l) + m \sum_{n=1}^{\infty} \ddot{q}_n(t)\,\sin(n\,\pi\,x/l) = f(x, t),$$

which gives

$$E\,I \sum_{n=1}^{\infty} q_n(t)\,(n\,\pi/l)^4 \int_0^l \sin(n\,\pi\,x/l)\,\sin(r\,\pi\,x/l)\,dx$$

$$+m \sum_{n=1}^{\infty} \ddot{q}_n(t) \int_0^l \sin(n\,\pi\,x/l)\,\sin(r\,\pi\,x/l)\,dx = \int_0^l f(x, t)\,\sin(r\,\pi\,x/l)\,dx,$$

by multiplication with $\sin(r\,\pi\,x/l)$ and integration over $(0, l)$, so that

$$E\,I\,q_r(t)\,(r\,\pi/l)^4\,\frac{l}{2} + m\,\ddot{q}_r(t)\,\frac{l}{2} = f_r(t), \quad r = 1, 2, \ldots, \text{ by orhtogonality,}$$

where $f_r(t) = \int_0^l f(x, t)\,\sin(r\,\pi\,x/l)\,dx$. The latter set of equations can be recast in the form

$$\ddot{q}_r(t) + \omega_r^2\,q_r(t) = \frac{2}{m\,l}\,f_r(t), \quad r = 1, 2, \ldots, \tag{5.17}$$

by using Eq. 5.9. The modal coordinates $\{q_n(t)\}$ satisfy differential equations of the type obtained for undamped SDOF systems in forced vibration (see Sects. 2.4.3 and 2.5.2). Their solutions require the initial conditions $\left(q_{r,0}, \dot{q}_{r,0}\right)$, $r = 1, 2, \ldots$, which can be obtained from the initial conditions $v_0(x) = v(x, 0)$ and $\dot{v}_0 = \dot{v}(x, 0)$ in the physical space. These conditions admit the representation

$$v_0(x) = \sum_{n=1}^{\infty} \varphi_n(x)\,q_{n,0} \quad \text{and} \quad \dot{v}_0(x) = \sum_{n=1}^{\infty} \varphi_n(x)\,\dot{q}_{n,0}$$

by Eq. 5.15, so that we have

$$\int_0^l v_0(x)\,\varphi_r(x)\,dx = \sum_{n=1}^{\infty}\left(\int_0^l \varphi_n(x)\,\varphi_r(x)\,dx\right) q_{n,0} = \frac{l}{2}\,q_{r,0} \quad \text{and}$$

$$\int_0^l \dot{v}_0(x)\,\varphi_r(x)\,dx = \sum_{n=1}^{\infty}\left(\int_0^l \varphi_n(x)\,\varphi_r(x)\,dx\right) \dot{q}_{n,0} = \frac{l}{2}\,\dot{q}_{r,0}, \quad r = 1, 2, \ldots, \tag{5.18}$$

for the system in Example 5.1 with modal shapes in Eq. 5.10.

The solutions of Eq. 5.17 with the initial conditions in Eq. 5.18, the modal shapes $\{\varphi_n(x)\}$, and the representation of $v(x,t)$ in Eq. 5.15 give the displacement function $v(x,t)$.

5.1.4 *Free Vibration*

The free vibration solution results from the previous section by setting $f = 0$. For completeness, we derive this solution directly. The representation of the displacement function in Eq. 5.14 and the equation of motion give

$$v(x,t) = \sum_{n=1}^{\infty} \sin\left(\frac{n\,\pi}{l}\,x\right)\big[A_n\,\cos(\omega_n\,t) + B_n\,\sin(\omega_n\,t)\big], \tag{5.19}$$

so that only the constants $\{A_n, B_n\}$ need to be determined. They result from the initial conditions $v(x,0) = v_0(x)$ and $\dot{v}(x,0) = \dot{v}_0(x)$, which, together with Eq. 5.19, give (see Example 5.1)

$$v_0(x) = \sum_{n=1}^{\infty} \sin\left(\frac{n\,\pi}{l}\,x\right) A_n \quad \text{and} \quad \dot{v}_0 = \sum_{n=1}^{\infty} \sin\left(\frac{n\,\pi}{l}\,x\right)(\omega_n\,B_n).$$

To find $\{A_n, B_n\}$, multiply the above equations by $\sin(m\,\pi\,x/l)$, integrate the resulting equations over the range $(0, l)$, and use the orthogonality property of Eq. 5.12. These operations give, e.g.,

$$\int_0^l v_0(x)\,\sin(m\,\pi\,x/l)\,dx = \sum_{n=1}^{\infty} A_n \int_0^l \sin(m\,\pi\,x/l)\,\sin(n\,\pi\,x/l)\,dx = \frac{l}{2}\,A_m,$$

so that we have

$$v(x,t) = \sum_{n=1}^{\infty}\left[\left(\frac{2}{l}\int_0^l v_0(x)\,\sin\left(\frac{n\,\pi}{l}\,x\right)dx\right)\cos(\omega_n\,t)\right]\sin\left(\frac{n\,\pi}{l}\,x\right) \tag{5.20}$$

for zero initial velocity $\dot{v}_0(x) = 0$.

5.2 Shear Beams

Shear beams provide simple models for the propagation of seismic waves from ruptures along faults through soil. Their displacement functions satisfy one-dimensional wave equations that are solved by following the approach of the previous section.

5.2.1 Physical System and Equations of Motion

The behavior of the shear beam resembles that of a deck of cards that cannot slide freely relative to each other. The infinitesimal element of length dx in Fig. 5.3 can be viewed as a small set of cards whose distortion requires a non-zero force increment, denoted in the figure by $\left(\partial Q(x,t)/\partial x\right) dx$.

Denote by $v(x,t)$, $Q(x,t)$, $\tau(x,t)$, and $f(x,t)$ the beam displacement, shear force, shear stress, and forcing function. It is assumed that the beam has constant cross-sectional area A, constant mass m per unit length, and constant (shear) modulus of elasticity G. The following differential equations define the behavior of shear beams subjected to arbitrary forcing functions $f(x,t)$ and can be found in, e.g., [3].

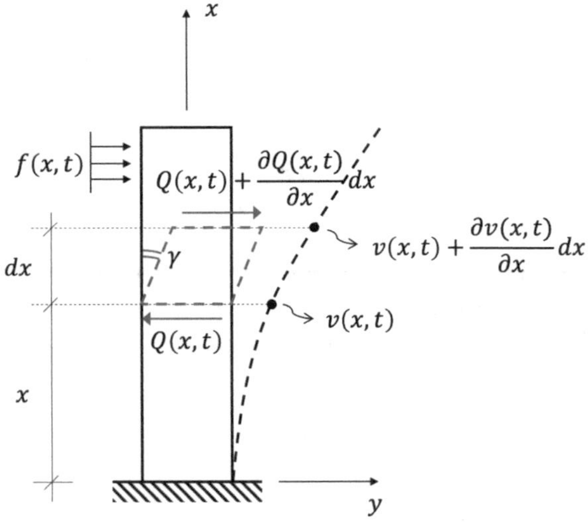

Fig. 5.3 Shear beam model

Distorsion of right angle: $\gamma(x,t) = \dfrac{\partial v(x,t)}{\partial x}$

Hooke's law: $\gamma(x,t) = \dfrac{\tau(x,t)}{G}$

Approximate force–stress relation: $\tau(x,t) = \dfrac{Q(x,t)}{A}$

Shear force-applied load relation: $\dfrac{\partial Q(x,t)}{\partial x} = -\text{“}f(x,t)\text{''},$

$$\text{where “}f(x,t)\text{''} = f(x,t) - m\,\ddot{v}(x,t). \tag{5.21}$$

The chain of equalities $\partial^2 v/\partial x^2 = \partial\gamma/\partial x = (1/G)\,\partial\tau/\partial x = (1/(G\,A))\,\partial Q/\partial x$ and the augmented force “f'' that includes inertia (see the latter relationship of Eq. 5.21) yield the beam equation

$$v'' = \frac{m}{G\,A}\,\ddot{v} - \frac{f}{G\,A}. \tag{5.22}$$

Its solution requires specifying initial and boundary conditions.

5.2.2 Modal Shapes and Frequencies

The homogeneous beam equation ($f = 0$),

$$\frac{\partial^2 v}{\partial t^2} = \frac{G\,A}{m}\frac{\partial^2 v}{\partial x^2} = c^2\,\frac{\partial^2 v}{\partial x^2}, \tag{5.23}$$

constitutes the one-dimensional wave equation, where $c^2 = G\,A/m$ denotes the square of the shear wave velocity. The method of separation of variables assumes that the solution has the form

$$v(x,t) = \varphi(x)\,q(t), \quad 0 \le x \le l, \quad t \ge 0, \tag{5.24}$$

where $\varphi(x)$ and $q(t)$ are functions of only space and only time arguments. This representation and Eq. 5.23 give

$$\varphi(x)\,\ddot{q}(t) = \frac{G\,A}{m}\,\varphi''(x)\,q(t) \quad \text{or, equivalently,} \quad \frac{\ddot{q}(t)}{q(t)} = \frac{G\,A}{m}\frac{\varphi''(x)}{\varphi(x)} = -\omega^2 \tag{5.25}$$

by using the arguments as for Eq. 5.5. The latter equalities of Eq. 5.25 give the following ordinary differential equations:

$$\ddot{q}(t) + \omega^2 q(t) = 0 \quad \text{(Initial value problem)} \quad \text{and}$$
$$\mathcal{D}[\varphi(x)] = \varphi''(x) + \rho^2 \varphi(x) = 0 \quad \text{(Boundary value problem)} \tag{5.26}$$

for $q(t)$ and $\varphi(x)$ with solutions

$$q(t) = A\,\cos(\omega\, t) + B\,\sin(\omega\, t)$$
$$\varphi(x) = C_1\,\cos(\rho\, x) + C_2\,\sin(\rho\, x), \tag{5.27}$$

where

$$\rho^2 = \frac{m\,\omega^2}{G\,A}. \tag{5.28}$$

The boundary conditions are used to find the constants C_1 and C_2 in the expression of $\varphi(x)$, construct the modal shapes, and find the modal frequencies. We illustrate this construction by the following example.

Example 5.2 Consider the shear beam in Fig. 5.3 with length l. The boundary conditions are $v(0,t) = 0$ and $Q(l,t) = 0$. The latter condition and Eq. 5.21 imply $\tau(x,t) = 0$ so that $\gamma(x,t) = \partial v(x,t)/\partial x = 0$ at $x = l$, which shows that the condition $v'(l,t) = 0$ can be substituted for $Q(l,t) = 0$.

The boundary condition $v(0,t) = \varphi(0)\,q(t) = 0$, $t \geq 0$, at $x = 0$ implies $\varphi(0) = 0$ so that $C_1 = 0$ and, as a result, $\varphi(x) = C_2\,\sin(\rho\,x)$. The boundary condition $v'(l,t) = \varphi'(l)\,q(t) = 0$, $t \geq 0$, at $x = l$ implies $\varphi(l) = C_2\,\sin(\rho\,l) = 0$. Since $C_2 = 0$ is not possible for non-zero initial conditions ($v(x,t) = 0$ at all times if $C_2 = 0$), we require $\sin(\rho\,l) = 0$. This equation has the (countable) infinite set of solutions

$$\rho_n\,l = \frac{(2\,n-1)\,\pi}{2} \quad \text{so that} \quad \frac{m\,l^2\,\omega_n^2}{G\,A} = \left(\frac{(2\,n-1)\,\pi}{2}\right)^2, \quad \text{which gives}$$
$$\omega_n = \frac{(2\,n-1)\,\pi}{2\,l}\sqrt{\frac{G\,A}{m}}, \quad n = 1, 2, \ldots. \tag{5.29}$$

The above values of ρ can also be found by considering the boundary conditions simultaneously. This approach yields homogeneous system of linear equations,

$$\begin{bmatrix} 1 & 0 \\ \rho\,\sin(\rho\,l) & \rho\,\cos(\rho\,l) \end{bmatrix} \begin{bmatrix} C_1 \\ C_2 \end{bmatrix} = \mathbf{0},$$

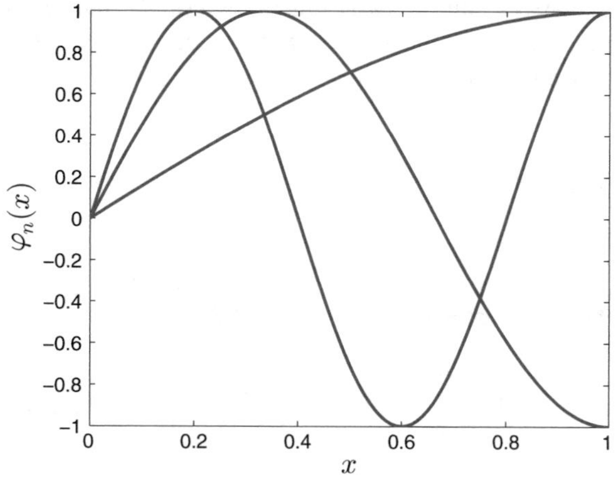

Fig. 5.4 First three modal shapes of a simply supported beam with $l = 1$

which admits non-trivial solutions if the determinant of its matrix vanishes, i.e., $\cos(\rho\, l) = 0$, which is the previous equation for ρ. The solutions $\varphi(x)$ corresponding to the above values of ρ are

$$\varphi_n(x) = \sin\left(\frac{(2\,n-1)\,\pi}{2\,l}\,x\right), \quad n = 1, 2, \ldots, \tag{5.30}$$

and are referred to as ***modal shapes*** corresponding to the ***modal frequencies*** $\{\omega_n\}$ in Eq. 5.29. The first three modal shapes are shown in Fig. 5.4. Note that the modes satisfy the boundary conditions, i.e., $\varphi_n(0) = 0$ and $\varphi'(l) = 0$, an expected property.

As mentioned previously, (1) the modal shapes are orthogonal in the sense

$$\int_0^l \varphi_n(x)\,\varphi_r(x)\,dx = \frac{l}{2}\,\delta_{nr}, \quad r = 1, 2, \ldots, \tag{5.31}$$

(2) the constant C_2 remains undetermined so that the function $\varphi(x)$ can be found up to a multiplicative constant, which is set unity in the expression of $\varphi_n(x)$, (3) the functions $\{q_n(t)\}$ are projections of the displacement function $v(x, t)$ on the basis functions $\{\varphi_n(x)\}$, (4) the functions $\varphi_n(x)\,q_n(t)$ satisfy Eq. 5.23 by construction, and (5) the modal shapes $\{\varphi_n(x)\}$ span the solution space so that

$$v(x, t) = \sum_{n=1}^{\infty} \varphi_n(x)\,q_n(t) = \sum_{n=1}^{\infty} \sin\left(\frac{(2\,n-1)\,\pi}{2\,l}\,x\right) q_n(t). \tag{5.32}$$

We reemphasize that the representation of the displacement function $v(x, t)$ in Eq. 5.32 is conceptually similar to that of the displacement vector of MDOF systems, a sum of basis functions $\{\varphi_n(x)\}$ that depend on the system mechanical properties, topology, and boundary conditions, which are weighted by the time-dependent functions $\{q_n(t)\}$.

5.2.3 Forced Vibration

The solution of the forced vibration problem of Eq. 5.22 and the representation of the displacement function in Eq. 5.32 give

$$\sum_{n=1}^{\infty} \varphi_n''(x)\, q_n(t) = \frac{m}{G\,A} \sum_{n=1}^{\infty} \varphi_n(x)\, \ddot{q}_n(t) - \frac{f(x,t)}{G\,A} \tag{5.33}$$

for the shear beam of Example 5.2. The multiplication of this equation by an arbitrary modal shape $\varphi_r(x)$ and the integration of the resulting equation over the beam domain $[0, l]$ give

$$-\rho_r^2 \frac{l}{2} q_r(t) = \frac{m}{G\,A} \frac{l}{2} \ddot{q}(t) - \frac{1}{G\,A} \int_0^l f(x,t)\, \varphi_r(x)\, dx$$

by using the orthogonality condition of Eq. 5.31. The latter equation can be recasted in the form

$$\ddot{q}_r(t) + \omega_r^2\, q_r(t) = f_r(t)/m, \quad p = 1, 2, \ldots, \tag{5.34}$$

where $f_r(t) = (2/l) \int_0^l f(x,t)\, \varphi_r(x)\, dx$. The solution of this equation results from our analysis of SDOF systems. For example, Eqs. 2.30 and 2.31 give

$$q_r(t) = A_r \cos(\omega_r\, t) + B_r \sin(\omega_r\, t) + \int_0^t h_r(t-u)\, f_r(u)\, du,$$

$$\text{where } h_r(t-u) = \frac{1}{m\,\omega_r} \sin\big(\omega_r\, t - u\big), \quad p = 1, 2, \ldots, \tag{5.35}$$

since the system has no damping. The constants result from the initial conditions $v(x,0) = v_0(x)$ and $\dot{v}(x,0) = \dot{v}_0(x)$ and the representation of $v(x,t)$ given by Eq. 5.32. We have

$$v_0(x) = \sum_{n=1}^{\infty} \varphi_n(x)\, q_n(0) = \sum_{n=1}^{\infty} \varphi_n(x)\, A_n \quad \text{and } \dot{v}_0(x) = \sum_{n=1}^{\infty} \varphi_n(x)\, \dot{q}_n(0) = \sum_{n=1}^{\infty} \varphi_n(x)\, \omega_n\, B_n,$$

so that

$$\int_0^l v_0(x)\,\varphi_r(x)\,dx = \sum_{n=1}^{\infty} A_n \int_0^l \varphi_n(x)\,\varphi_r(x)\,dx = \frac{l}{2}\,A_r \quad \text{and}$$

$$\int_0^l \dot{v}_0(x)\,\varphi_r(x)\,dx = \sum_{n=1}^{\infty} \omega_n\,B_n \int_0^l \varphi_n(x)\,\varphi_r(x)\,dx = \frac{l}{2}\,\omega_r\,B_r, \quad p = 1, 2, \ldots, \tag{5.36}$$

by using the orthogonality condition of Eq. 5.31. The general solution of Eq. 5.34 is

$$q_r(t) = \left(\frac{2}{l}\int_0^l v_0(x)\,\varphi_r(x)\,dx\right)\cos(\omega_r\,t) + \left(\frac{2}{\omega_r\,l}\int_0^l \dot{v}_0(x)\,\varphi_r(x)\,dx\right)\sin(\omega_r\,t)$$
$$+ \int_0^t h_r(t-u)\,f_r(u)\,du, \tag{5.37}$$

so that

$$v(x,t) = \sum_{n=1}^{\infty} \sin\left(\frac{(2n-1)\,\pi}{2l}\,x\right)\left[\left(\frac{2}{l}\int_0^l v_0(x)\,\varphi_n(x)\,dx\right)\cos(\omega_n\,t)\right.$$
$$\left. + \left(\frac{2}{\omega_n\,l}\int_0^l \dot{v}_0(x)\,\varphi_n(x)\,dx\right)\sin(\omega_n\,t) + \int_0^t h_n(t-u)\,f_n(u)\,du\right]. \tag{5.38}$$

5.2.4 Free Vibration

The free vibration solution can be obtained from the previous subsection by setting $f(x,t) = 0$. For completeness, we outline the steps for finding the free vibration solution directly. The results are for the shear beam of Example 4.60.

Previous arguments on Eqs. 5.33 to 5.34 with $f(x,t) = 0$ give

$$\ddot{q}_n(t) + \omega_r^2\,q_n(t) = 0, \quad n = 1, 2, \ldots, \tag{5.39}$$

which is satisfied by

$$q_n(t) = A_n\,\cos(\omega_n\,t) + B_n\,\sin(\omega_n\,t), \quad n = 1, 2, \ldots, \tag{5.40}$$

where the constants $\{A_n\}$ and $\{B_n\}$ result from the initial conditions $v(x,0) = v_0(x)$ and $\dot{v}(x,0) = \dot{v}_0(x)$ by using the representation of $v(x,t)$ given by Eq. 5.32 and the orthogonality of the modal shapes. The free vibration solution is

$$v(x,t) = \sum_{n=1}^{\infty} \sin\left(\frac{(2n-1)\pi}{2l}x\right)\left[\left(\frac{2}{l}\int_0^l v_0(x)\,\varphi_r(x)\,dx\right)\cos(\omega_n t)\right.$$

$$\left. + \left(\frac{2}{\omega_n l}\int_0^l \dot{v}_0(x)\,\varphi_r(x)\,dx\right)\sin(\omega_n t)\right]. \tag{5.41}$$

We conclude with the observation that the infinite series representations of the solutions $v(x,t)$, e.g., the series of Eqs. 5.20, 5.32, and 5.38, have to be truncated for numerical calculations. The accuracy of the resulting calculations depends on the numbers of terms retained from the infinite series of $v(x,t)$ and the rate of convergence of these series.

5.3 Problems

Problem 5.1 The continuous beam in Example 5.1 is subjected to the initial conditions $v_0(x) = \varphi_n(x) = \sin\left(n\pi x/l\right)$ and $\dot{v}_0(x) = 0$. Find the free vibration solution of the beam.

Problem 5.2 The continuous beam in Example 5.1 is subjected to the harmonic forcing function $f(x,t) = q\,\sin(\nu t)$ with frequency $\nu > 0$ and amplitude $q > 0$. Find the forced vibration solution of the beam.

Problem 5.3 Consider a cantilever of length l with constant stiffness EI and constant mass per unit length m. Find the first five modal frequencies and shapes and plot the modal shapes. View the cantilever as a flexural beam. Assume $l = 1$, $EI = 1$, and $m = 1$.
Hint: The following steps can be used for solution: (1) Use the general expression of $\varphi(x)$ in Eq. 5.8. (2) Impose the boundary conditions at the fixed end, and use resulting relationships to eliminate two constants from the expression of $\varphi(x)$. (3) Impose the boundary conditions at the free end to obtain a linear homogeneous system of equations in the remaining two constants in the expression of $\varphi(x)$. Since the trivial solution is not possible, set the determinant of this system zero. The resulting equation is

$$\cos(\beta l)\cosh(\beta l) = -1, \quad \text{or} \quad \cos(\beta l) + \frac{1}{\cosh(\beta l)} = 0,$$

where β is defined in Eq. 5.8. The first five roots of the equation give the first five modal frequencies. (4) Use any of the 2 equations imposing the boundary conditions at the free end to construct modal shapes.

Problem 5.4 A simple support is added to the cantilever of the previous problem. Find the first five modal frequencies and shapes. Plot the modal shapes.

Problem 5.5 The shear beam in Example 5.2 is subjected to the harmonic forcing function $f(x, t) = q \sin(\nu t)$ with frequency $\nu > 0$ and amplitude $q > 0$. Find the forced vibration solution of the beam. Assume zero initial conditions.

Problem 5.6 A simple support is added to the cantilever of the shear beam of the previous problem. Find the first five modal frequencies and shapes. Plot the modal shapes.

References

1. B. Friedman, *Principles and Techniques of Applied Mathematics* (John Wiley & Sons Inc., New York, 1956)
2. I. Gohberg, S. Goldberg, *Basic Operator Theory* (Birkhäuser, Boston, 1980)
3. E.P. Popov, *Engineering Mechanics of Solids* (Prentice Hall, New York, 1998)

Appendix A
Taylor Series

Consider a real-valued function $f(t)$ whose first n derivatives are continuous on an interval $[a, b]$ and its derivative of order $(n + 1)$ exists on this interval. Then, for each t in $[a, b]$, there is c between a and t such that

$$f(t) = f(a) + \sum_{r=1}^{n} \frac{(t-a)^r}{r!} f^{(r)}(a) + R_n(c, t), \tag{A.1}$$

where

$$R_n(c, t) = \frac{(t-a)^{n+1}}{(n+1)!} f^{(n+1)}(c) \tag{A.2}$$

is the remainder of the representation of $f(t)$, and $f^{(r)}(a)$ denotes the rth derivative $d^r f(t)/dt^r$ of $f(t)$ at $t = a$ [2] (Theorem 21.1).

This implies that the error of truncated series

$$f_n(t) = f(a) + \sum_{r=1}^{n} \frac{(t-a)^r}{r!} f^{(r)}(a) \tag{A.3}$$

is of the order $R_n(c, t)$, i.e., order $O\big((t-a)^{n+1}\big)$. We use this observation to construct finite difference representation of derivatives and other approximations.

The Taylor theorem can be used to show the validity of the L'Hôpital rule. Suppose that (1) the ratio $f_1(t)/f_2(t)$ of the functions $f_1(t)$ and $f_2(t)$ is indeterminate at $t = t_0$, e.g., $f_1(t_0) = f_2(t_0) = 0$, which results in $0/0$, (2) the functions f_1 and f_2 are differentiable, and (3) the derivative of f_2 is not zero at t_0, i.e., $f_2'(t_0) \neq 0$. The Taylor series expansions of these functions around t_0 give

M. D. Grigoriu, *Linear Dynamical Systems*,
https://doi.org/10.1007/978-3-030-64552-6

$$\begin{aligned}\frac{f_1(t)}{f_2(t)} &= \frac{f_1(t_0) + f_1'(t_0)\,(t-t_0) + O\big(|t-t_0|^2\big)}{f_2(t_0) + f_2'(t_0)\,(t-t_0) + O\big(|t-t_0|^2\big)} \\ &= \frac{f_1'(t_0)\,(t-t_0) + O\big(|t-t_0|^2\big)}{f_2'(t_0)\,(t-t_0) + O\big(|t-t_0|^2\big)} = \frac{f_1'(t_0) + O\big(|t-t_0|\big)}{f_2'(t_0) + O\big(|t-t_0|\big)} \to \frac{f_1'(t_0)}{f_2'(t_0)}, \\ &\quad \text{as } t \to t_0. \end{aligned} \tag{A.4}$$

The L'Hôpilal rule can be used to find the limit of a broad range of indeterminate forms such as $0/0$, ∞/∞, and $0 \times \infty$. For example, the ratio of $f_1(t) = t^2 + t - 6$ to $f_2(t) = t^2 - 4$ is $0/0$ at $t_0 = 2$, and the L'Hôpilal rule gives

$$\lim_{t \to 2} \frac{t^2 + t - 6}{t^2 - 4} = \lim_{t \to 2} \frac{2t + 1}{2t} = \frac{5}{4}.$$

However, the rule fails in some cases. For example, the rule fails for $f_1(t) = t + \cos(t)$ and $f_2(t) = t$ and $t_0 = \infty$ although the functions are differentiable and $f_2'(t) = 1$. The L'Hôpilal rule gives

$$\lim_{t \to \infty} \frac{t + \cos(t)}{t} = \lim_{t \to \infty} \frac{1 - \sin(t)}{1},$$

which does not have limit as $t \to \infty$. Yet,

$$\lim_{t \to \infty} \frac{t + \cos(t)}{t} = \lim_{t \to \infty} \left(1 + \frac{\cos(t)}{t}\right) = 1.$$

Appendix B
Linear Differential Equations

Consider the homogeneous linear differential equation of order n with constant coefficients

$$a_0\,x(t) + a_1\,x'(t) + a_2\,x''(t) + \cdots + a_n\,x^{(n)}(t) = 0, \quad t \geq 0, \tag{B.1}$$

where the coefficients $\{a_i\}$ are real numbers and $x^{(n)}$ denotes derivative of order n. The exponential function $\exp(\lambda\,t)$ satisfies Eq. B.1 if λ is such that

$$\left(a_0 + a_1\,\lambda + a_2\,\lambda^2 + \cdots + a_n\,\lambda^n\right) e^{\lambda\,t} = 0, \quad t \geq 0, \tag{B.2}$$

which implies

$$p(\lambda) = a_0 + a_1\,\lambda + a_2\,\lambda^2 + \cdots + a_n\,\lambda^n = 0 \tag{B.3}$$

since $e^{\lambda\,t} \neq 0$ for finite $\lambda\,t$. The condition $p(\lambda) = 0$ of Eq. B.3 is referred to as the ***characteristic equation***. Since $p(\lambda)$ is a polynomial of the nth degree, it has n roots, which can be real or complex even for real-valued coefficients $\{a_i\}$. There are two cases.

The first is that of distinct solutions $\lambda_1, \lambda_2, \cdots, \lambda_n$ of $p(\lambda) = 0$. Then, the exponential functions $\{\exp(\lambda_i\,t)\}$ (1) satisfy Eq. B.1 by construction, (2) are linearly independent, and (3) constitute a basis for the solution of Eq. B.1 [4]. This means that the solution of Eq. B.1 can be viewed as an element of the linear space spanned by these exponential functions so that

$$x(t) = \sum_{i=1}^{n} \alpha_i\,e^{\lambda_i\,t}, \quad t \geq 0, \tag{B.4}$$

where the coefficients $\{\alpha_i\}$ result from initial conditions.

M. D. Grigoriu, *Linear Dynamical Systems*,
https://doi.org/10.1007/978-3-030-64552-6

The second is that in which one or more solutions of $p(\lambda) = 0$ are multiple, in which case the set of distinct functions $\{\exp(\lambda_i\, t)\}$ does not span the solution space. For example, suppose λ_1 is multiple of order $m < n$, and the other roots are simple. This means that Eq. B.3 has the form $p(\lambda) = \left(\lambda - \lambda_1\right)^m \tilde{p}(\lambda) = 0$, where $\tilde{p}(\lambda)$ is a polynomial of degree, $n - m$. It can be shown that the solution of Eq. B.1 is an element of the linear space spanned by the functions

$$e^{\lambda_1 t}, t\, e^{\lambda_1 t}, \cdots, t^{m-1}\, e^{\lambda_1 t}, e^{\lambda_2 t}, \cdots, e^{\lambda_{n-m+1} t},$$

for our illustration so that the solution has the form [4]

$$x(t) = \sum_{i=1}^{m} \alpha_i\, t^{i-1}\, e^{\lambda_i t} + \sum_{i=2}^{n-m+1} \alpha_{m+i-1}\, e^{\lambda_i t}, \quad t \geq 0. \tag{B.5}$$

Similar solutions result for the case in which two or more roots are multiple.

Consider now the inhomogeneous version of Eq. B.1, i.e., the equation

$$a_0\, x(t) + a_1\, x'(t) + a_2\, x''(t) + \cdots + a_n\, x^{(n)}(t) = f(t), \quad t \geq 0, \tag{B.6}$$

where $f(t)$ is a specified function. Generally, the method of the ***variation of constants*** is used to construct particular solutions of this equation [1] (Sect. 2.4). We have seen that the general solution of the homogeneous equation has the form

$$x(t) = \sum_{i=1}^{n} \alpha_i\, \beta_i(t), \quad t \geq 0, \tag{B.7}$$

where the basis functions $\{\beta_i(t)\}$ are exponential functions or mixtures of exponential and exponentials scaled by polynomials for simple and multiple roots of the characteristic equation. The method assumes that the particular solutions $x_p(t)$ of Eq. B.6 have the form of Eq. B.7, i.e.,

$$x_p(t) = \sum_{i=1}^{n} \alpha_i(t)\, \beta_i(t), \quad t \geq 0, \tag{B.8}$$

but the coefficients $\{\alpha_i\}$ are unknown functions $\{\alpha_i(t)\}$ of time, which are required to satisfy the conditions

$$\sum_{i=1}^{n} \alpha_i'(t)\, \beta_i^{(k)}(t) = 0, \quad k = 0, 1, \ldots, n-2. \tag{B.9}$$

Under these conditions, we have $x_p^{(j)}(t) = \sum_{i=1}^{n} \alpha_i(t)\, \beta_i^{(j)}(t)$, $j = 0, 1, \ldots, n-1$, and $x_r^{(n)}(t) = \sum_{i=1}^{n} \left[\alpha_i'(t)\, \beta_i^{(n-1)}(t) + \alpha_i(t)\, \beta_i^{(n)}(t)\right]$. These expressions of the derivatives of $x_p(t)$ and the differential equation give

$$\sum_{i=1}^{n} \alpha_i'(t)\,\beta_i^{(n-1)} = f(t). \tag{B.10}$$

The conditions of Eqs. B.9 and B.10 define a linear system of equations for the first derivatives of the unknown functions $\{\alpha_i(t)\}$ of the form $\alpha_i'(t) = g_i(t)$, $i = 1, \ldots, n$, where the functions $g_i(t)$ depend on the basis functions and the forcing function.

Example B.1 Suppose $x(t)$ is the displacement of the SDOF system defined by Eq. 2.8 with damping ratio $0 < \zeta < 1$. The basis functions are $\beta_1(t) = \exp(-\zeta\,\omega\,t)\,\cos(\omega_d\,t)$ and $\beta_2(t) = \exp(-\zeta\,\omega\,t)\,\sin(\omega_d\,t)$. The differential operator of the defining equation of $x(t)$ is $\mathcal{D} = d^2/dt^2 + 2\,\zeta\,\omega\,d/dt + \omega^2$, and the order of this equation is $n = 2$. From Eq. B.9, we have $\alpha_1'(t)\,\beta_1(t) + \alpha_2'(t)\,\beta_2(t) = 0$, so that $\dot{x}_p(t) = \alpha_1(t)\,\beta_1'(t) + \alpha_2(t)\,\beta_2'(t)$, $\ddot{x}_p(t) = \alpha_1'(t)\,\beta_1'(t) + \alpha_2'(t)\,\beta_2'(t) + \alpha_1(t)\,\beta_1''(t) + \alpha_2(t)\,\beta_2''(t)$, and

$$\begin{aligned}
\mathcal{D}[x_p(t)] &= \alpha_1(t)\,\mathcal{D}[\beta_1(t)] + \alpha_2(t)\,\mathcal{D}[\beta_2(t)] + \alpha_1'(t)\,\beta_1'(t) + \alpha_2'(t)\,\beta_2'(t) \\
&= \alpha_1'(t)\,\beta_1'(t) + \alpha_2'(t)\,\beta_2'(t),
\end{aligned}$$

where the latter equality holds since $\{\beta_i(t)\}$ satisfies the homogeneous equation. Since $x_p(t)$ must satisfy the inhomogeneous equation, we have $\alpha_1'(t)\,\beta_1'(t) + \alpha_2'(t)\,\beta_2'(t) = f(t)$. The resulting conditions for $\{\alpha_i'(t)\}$ are (see also Eqs. B.9 and B.10)

$$\begin{cases} \alpha_1'(t)\,\beta_1(t) + \alpha_2'(t)\,\beta_2(t) = 0 \\ \alpha_1'(t)\,\beta_1'(t) + \alpha_2'(t)\,\beta_2'(t) = f(t) \end{cases}$$

whose solutions are $\alpha_i'(t) = g_i(t)$, $i = 1, 2$, where

$$\begin{aligned}
g_1(t) &= -\beta_2(t)\,f(t)/\big(\beta_1(t)\,\beta_2'(t) - \beta_1'(t)\,\beta_2(t)\big) \quad \text{and} \\
g_2(t) &= \beta_2 1(t)\,f(t)/\big(\beta_1(t)\,\beta_2'(t) - \beta_1'(t)\,\beta_2(t)\big).
\end{aligned}$$

The particular solution has the form $x_p(t) = \big(\int g_1(t)\,dt\big)\,\beta_1(t) + \big(\int g_2(t)\,dt\big)\,\beta_2(t)$.

We have not used the method of the variation of constants to construct particular solutions for the forced vibration of SDOF systems for two reasons. First, the method is rather abstract. Second, particular solutions can be constructed by direct observations in simple cases, e.g., constant and harmonic forces, and, for arbitrary forcing functions, from dynamical responses of SDOF systems to elementary actions (see Duhamel's integral in Sect. 2.4.4).

The following example deals with a differential equation, which is essential for the analysis MDOF systems with non-proportional damping. It shows that the solution of this equation can be obtained by elementary arguments.

Example B.2 Consider the differential equation

$$\dot{q}(t) = \lambda\, q(t) + h(t), \quad t \geq 0, \tag{B.11}$$

with the initial condition $q(0) = q_0$.

Assume first that λ and $h(t)$ are real. The homogeneous equation $\dot{q}(t) = \lambda\, q(t)$ can be recast in the form $dq(t)/q(t) = \lambda\, dt$ so that $\ln\big(q(t)\big)\big|_0^t = \lambda\, t$, which gives $q(t) = c\, \exp(\lambda\, t)$, where c is an arbitrary constant. We claim that the solution of the inhomogeneous equation has the form

$$q(t) = q_0\, e^{\lambda\, t} + \int_0^t e^{\lambda\,(t-s)}\, h(s)\, ds, \quad t \geq 0. \tag{B.12}$$

We only show that $q(t)$ in Eq. B.12 satisfies Eq. B.11. The derivative of the first term in the expression of $q(t)$ is $q_0\, \lambda\, \exp(\lambda\, t)$. The derivative of the second term

$$I(t) = \int_0^t e^{\lambda\,(t-s)}\, h(s)\, ds, \quad t \geq 0,$$

results by taking the limit of $\big(I(t+\Delta t) - I(t)\big)/\Delta t$ as $\Delta t \to 0$. With the notation $g(t,s) = \exp\big(\lambda\,(t-s)\big)\, h(s))$, we have

$$\begin{aligned}\frac{I(t+\Delta t) - I(t)}{\Delta t} &= \frac{1}{\Delta t}\left[\int_0^{t+\Delta t} g(t+\Delta t, s)\, ds - \int_0^t g(t,s)\, ds\right]\\ &= \int_0^t \frac{g(t+\Delta t, s) - g(t,s)}{\Delta t}\, ds + \frac{1}{\Delta t}\int_t^{t+\Delta t} g(t+\Delta t, s)\, ds\\ &\to \int_0^t \frac{\partial g(t,s)}{\partial t}\, ds + g(t,t), \quad \text{as } \Delta t \to 0,\end{aligned}$$

so that $\int_0^t \lambda\, \exp\big(\lambda\,(t-s)\big)\, h(s)\, ds + h(t)$ is the time derivative of the second term in the expression of $q(t)$ and

$$\begin{aligned}\dot{q}(t) &= q_0\, \lambda\, e^{\lambda\, t} + \int_0^t \lambda\, e^{\lambda\,(t-s)}\, h(s)\, ds + h(t) = \lambda\left[q_0\, e^{\lambda\, t} + \int_0^t e^{\lambda\,(t-s)}\, h(s)\, ds\right]\\ &= \lambda\, q(t) + h(t).\end{aligned}$$

We conclude that $q(t)$ in Eq. B.12 is a solution of Eq. B.11. It is the solution of this equation by uniqueness [4].

The extension to complex-valued λ and/or initial condition q_0 is simple since the resulting complex-valued solution $q(t)$ is defined on the real line so that it can be viewed as a two-dimensional vector whose components are its real and imaginary parts. Alternatively, we note that the above operations are also valid for complex-valued functions defined on the real line such as $q(t)$ so that $q(t)$ in Eq. B.12 solves Eq. B.11 for complex-valued λ and/or initial condition q_0.

Appendix C
Linear or Vector Spaces

This appendix presents properties of linear or vector spaces which are relevant to the analysis of MDOF and continuous systems. It defines vector spaces, inner products on these spaces, basis, and linear independence. The discussion is kept simple as it is intended to develop intuition on these spaces and provide useful facts for dynamical analysis. A rigorous treatment of these topics can be found in, e.g., [3].

Definition C.1 A ***linear*** or ***vector space*** $\mathcal{V}$ is a set that is closed to addition and multiplication by scalar, i.e., if $\mathbf{x}$ and $\mathbf{y}$ are in $\mathcal{V}$, then $\mathbf{x} + \mathbf{y}$ and $\alpha\,\mathbf{x}$ are in $\mathcal{V}$, where the scalar α is assumed to be a real number in our considerations. The operations $\mathbf{x} + \mathbf{y}$ and $\alpha\,\mathbf{x}$ are performed component-by-component, i.e., the components of $\mathbf{x} + \mathbf{y}$ and $\alpha\,\mathbf{x}$ are $\{x_k + y_k\}$ and $\{\alpha\, x_k\}$, where $\{x_k\}$ and $\{y_k\}$, $k = 1, \ldots, n$, denote the components of $\mathbf{x}$ and $\mathbf{y}$.

Example C.1 The physical three-dimensional space, which is denoted by $\mathbb{R}^3$, is a vector space. The elements $\mathbf{x}$ of this space can be represented by

$$\mathbf{x} = x_1\,\mathbf{i} + x_2\,\mathbf{j} + x_3\,\mathbf{k}, \tag{C.1}$$

where the components x_1, x_2, and x_3 of $\mathbf{x}$ are the projections of this vector on the unit vectors $\mathbf{i}$, $\mathbf{j}$, and $\mathbf{k}$ of the system of coordinates. If $\mathbf{y}$ is another three-dimensional vector with components y_1, y_2, and y_3, then $\mathbf{x}+\mathbf{y}$ is also a three-dimensional vector with components $x_1 + y_1$, $x_2 + y_2$, and $x_3 + y_3$. Also, $\alpha\,\mathbf{x}$ is a three-dimensional vector with components $\alpha\,x_1$, $\alpha\,x_2$, and $\alpha\,x_3$.

The following two examples are extensions of the physical space, a three-dimensional vector space, to vector spaces $\mathcal{V}$ with finite and infinite numbers of components.

Example C.2 Suppose that the vectors $\mathbf{x}$ and $\mathbf{y}$ of the previous example belong to an n-dimensional vector space $\mathbb{R}^n$, also referred to as the n-dimensional ***Euclidean space***. Denote, as previously, the unit vectors of this space by $\mathbf{i}_k$, $k = 1, \ldots, n$. In

M. D. Grigoriu, *Linear Dynamical Systems*,
https://doi.org/10.1007/978-3-030-64552-6

this system of coordinates, **x** and **y** admit the representations

$$\mathbf{x} = \sum_{k=1}^{n} x_k\, \mathbf{i}_k \quad \text{and} \quad \mathbf{y} = \sum_{k=1}^{n} y_k\, \mathbf{i}_k, \tag{C.2}$$

where $\{x_k\}$ and $\{y_k\}$, $k = 1, \ldots, n$, denote the components of **x** and **y**, i.e., the projections of these vectors on the unit vectors $\{\mathbf{i}_k\}$. The vector addition and multiplication by scalar is performed, as previously, component-by-components, i.e.,

$$\mathbf{x} + \mathbf{y} = \sum_{k=1}^{n} \big(x_k + y_k\big)\, \mathbf{i}_k$$

$$\alpha\, \mathbf{x} = \sum_{k=1}^{n} \big(\alpha\, x_k\big)\, \mathbf{i}_k. \tag{C.3}$$

Example C.3 Suppose now that **x** and **y** of the previous example are elements, f and g, of a space of functions, e.g., the space of periodic continuous functions defined on a time interval $[0, \tau]$. We have seen in Sect. 2.6.2 that the elements of this space can be represented by Fourier series as infinite sums of $\cos(\nu_k\, t)$ and $\sin(\nu_k\, t)$, $t \in [0, \tau]$, where $\nu_1 = 2\,\pi/\tau$ and $\nu_k = k\, \nu_1$. The elements f and g of this space of functions admit the representations

$$f(t) = \frac{a_0}{2} + \sum_{k=1}^{\infty} \big(a_k\, \cos(\nu_k\, t) + b_k\, \sin(\nu_k\, t)\big)$$

$$g(t) = \frac{a_0'}{2} + \sum_{k=1}^{\infty} \big(a_k'\, \cos(\nu_k\, t) + b_k'\, \sin(\nu_k\, t)\big), \tag{C.4}$$

where the coefficients $\{a_k\}$, $\{b_k\}$ $\{a_k'\}$, and $\{b_k'\}$ are given in Eq. 2.53. The vector addition and multiplication by scalars is performed, as previously, "component-by-component," i.e.,

$$f(t) + g(t) = \frac{a_0 + a_0'}{2} + \sum_{k=1}^{\infty} \big((a_k + a_k')\, \cos(\nu_k\, t) + (b_k + b_k')\, \sin(\nu_k\, t)\big)$$

$$\alpha\, f(t) = \frac{\alpha\, a_0}{2} + \sum_{k=1}^{\infty} \big(\alpha\, a_k\, \cos(\nu_k\, t) + \alpha\, b_k\, \sin(\nu_k\, t)\big), \tag{C.5}$$

Definition C.2 The ***inner product*** on a vector space $\mathcal{V}$ is an operation, which associates with any pair $\mathbf{x}$ and $\mathbf{y}$ of elements of $\mathcal{V}$ a scalar $\langle \mathbf{x}, \mathbf{y} \rangle$ (real or complex) and has the following properties: (1) $\langle \mathbf{x}, \mathbf{x} \rangle > 0$ unless $\mathbf{x} = \mathbf{0}$ in which case $\langle \mathbf{x}, \mathbf{x} \rangle = 0$, (2) $\langle \mathbf{x}, \mathbf{y} \rangle = \langle \mathbf{y}, \mathbf{x} \rangle^*$, and (3) $\alpha \langle \mathbf{x}, \mathbf{y} \rangle = \langle \alpha \mathbf{x}, \mathbf{y} \rangle$ and $\langle \mathbf{x} + \mathbf{z}, \mathbf{y} \rangle = \langle \mathbf{x}, \mathbf{y} \rangle + \langle \mathbf{z}, \mathbf{y} \rangle$, where $\mathbf{z}$ is a vector in $\mathcal{V}$ and the symbol * denotes complex conjugate.

Definition C.3 The inner product in the Euclidian space $\mathbb{R}^n$ is

$$\langle \mathbf{x}, \mathbf{y} \rangle = \sum_{i=1}^{n} x_i \, y_i, \quad \mathbf{x}, \mathbf{y} \in \mathbb{R}^n, \tag{C.6}$$

where $\{x_i\}$ and $\{y_i\}$ denote the components of $\mathbf{x}$ and $\mathbf{y}$. If $\langle \mathbf{x}, \mathbf{y} \rangle = 0$, the vectors $\mathbf{x}$ and $\mathbf{y}$ in $\mathcal{V}$ are said to be ***orthogonal***.

Example C.4 The operations with vectors of the physical space considered in Example C.1 satisfy the properties of the inner product since (see Eq. C.6)

$$\langle \mathbf{x}, \mathbf{y} \rangle = x_1 \, y_1 + x_2 \, y_2 + x_3 \, y_3, \tag{C.7}$$

and the vector addition and multiplication by scalars is performed component-by-component. Note that $\langle \mathbf{x}, \mathbf{x} \rangle = x_1^2 + x_2^2 + x_3^2$ is the square of the length of $\mathbf{x}$, which is zero if and only if $\mathbf{x} = \mathbf{0}$ is the null vector.

Example C.5 Similar arguments hold for the finite-dimensional vector space of Example C.2 since $\langle \mathbf{x}, \mathbf{y} \rangle = \sum_{i=1}^{n} x_i \, y_i$, and the vector addition and multiplication by scalars is performed component-by-component. As mentioned previously, the square of the length of $\mathbf{x}$ is $\sum_{i=1}^{n} x_i^2$ so that $\langle \mathbf{x}, \mathbf{x} \rangle = 0$ if and only if $\mathbf{x} = \mathbf{0}$.

Example C.6 The values of the elements f and g of the vector space of functions in Example C.3 at arbitrary t can be viewed as "components" of these infinite-dimensional vectors. The extension of Eq. C.6 to the continuous case yields the inner product

$$\langle f, g \rangle = \int_0^{\tau} f(t) \, g(t) \, dt. \tag{C.8}$$

That this definition satisfies the properties of the inner product follows from the operations defined by Eq. C.5. Note also that the square of the "length" of f is $\langle f, f \rangle = \int_0^{\tau} f(t)^2 \, dt$ so that it is zero if only if $f(t) = 0$, $t \in [0, \tau]$.

Definition C.4 A set of vectors $\mathbf{x}_r$, $r = 1, \ldots, p$, of $\mathbb{R}^n$ is ***linearly independent*** if

$$\sum_{r=1}^{p} \alpha_r \, \mathbf{x}_r = \mathbf{0} \quad \text{implies } \alpha_r = 0, \quad r = 1, \ldots, p, \tag{C.9}$$

i.e., the equation $\sum_{r=1}^{p} \alpha_r \mathbf{x}_r = 0$ admits only the trivial solution. If this does not hold, the vectors are said to be ***linearly dependent***.

Note

1. A set of vectors $\mathbf{x}_r$, $r = 1, \ldots, p$, of $\mathbb{R}^n$ containing the zero vector is linearly dependent since, if $\mathbf{x}_p = \mathbf{0}$, $\sum_{r=1}^{p} \alpha_r \mathbf{x}_r = \mathbf{0}$ holds for $\alpha_1 = \cdots = \alpha_{p-1} = 0$ and $\alpha_p \neq 0$.
2. If $p > n$, the set of vectors $\mathbf{x}_r$, $r = 1, \ldots, p$, is linearly dependent. Let $\mathbf{A}$ be an (n, p)-matrix whose columns are the vectors $\mathbf{x}_r$. The system of equations $\mathbf{A}\,\alpha = \sum_{r=1}^{p} \alpha_r \mathbf{x}_r = \mathbf{0}$ has more variables than equations so that it admits non-trivial solutions, i.e., $\sum_{r=1}^{p} \alpha_r \mathbf{x}_r = \mathbf{0}$ does not imply $\alpha_r = 0$ for all $r = 1, \ldots, p$.
3. A direct consequence of the previous statement is that n is the maximum number of linearly independent vectors in $\mathbb{R}^n$.

Example C.7 The unit vectors **i**, **j**, and **k** are linearly independent since

$$\alpha_1 \mathbf{i} + \alpha_2 \mathbf{j} + \alpha_3 \mathbf{k} = \begin{bmatrix} \alpha_1 \\ \alpha_2 \\ \alpha_3 \end{bmatrix} = \mathbf{0} \quad \text{if and only if } \alpha_1 = \alpha_2 = \alpha_3 = 0.$$

Similarly, the "vectors" $\cos(\nu_k t)$ and $\sin(\nu_k t)$ are linearly independent since $f(t) = 0$ at all times t in $[0, \tau]$, if and only if the coefficients $\{a_k\}$ and $\{b_k\}$ of the representation in Eq. C.4 are zero.

Definition C.5 A subset $\mathcal{B} = \{\mathbf{b}_1, \cdots, \mathbf{b}_m\}$ of a vector space $\mathcal{V}$ is a ***basis*** of $\mathcal{V}$ if (1) the elements of every finite subset of $\mathcal{B}$ are linearly independent and (2) the elements $\mathbf{x}$ of $\mathcal{V}$ can be represented uniquely by $\mathbf{x} = \sum_{r=1}^{m} \alpha_r \mathbf{b}_r$, i.e., $\mathcal{B}$ spans the vector space $\mathcal{V}$.

Example C.8 The unit vectors **i**, **j**, and **k** of the physical space $\mathbb{R}^3$ define a basis for this space since they are linearly independent and they span the space. The latter statement follows from the observation that **i**, **j**, **k**, and any vector $\mathbf{x} \in \mathbb{R}^3$ are dependent, so that $\alpha_1' \mathbf{i} + \alpha_2' \mathbf{j} + \alpha_3' \mathbf{k} + \alpha_4' \mathbf{x} = \mathbf{0}$ holds with non-zero scalars, so that $\mathbf{x}$ is a linear form of **i**, **j**, and **k**. The projections of $\mathbf{x}$ on the unit vectors are $\langle \mathbf{x}, \mathbf{i} \rangle$, $\langle \mathbf{x}, \mathbf{j} \rangle$, and $\langle \mathbf{x}, \mathbf{k} \rangle$ so that this vector admits the representation $\mathbf{x} = \langle \mathbf{x}, \mathbf{i} \rangle\, \mathbf{i} + \langle \mathbf{x}, \mathbf{j} \rangle\, \mathbf{j} + \langle \mathbf{x}, \mathbf{k} \rangle\, \mathbf{k}$.

Example C.9 Similar considerations hold for the n-dimensional Euclidian space $\mathbb{R}^n$. Its unit vectors and other linearly independent vectors, e.g., the eigenvectors of an (n, n)-matrix $\mathbf{a}$, define basis of this space.

Suppose first that $\mathbf{a}$ is an (n, n)-symmetric real-valued matrix, and denote by $\mathbf{x}_1, \ldots, \mathbf{x}_n$ its eigenvectors. These eigenvectors and an arbitrary vector $\mathbf{x}$ of $\mathbb{R}^n$ are linearly dependent (see item 3 following Definition C.4) so that $\sum_{r=1}^{n} \alpha_r \mathbf{x}_r + \alpha_{n+1} \mathbf{x} = \mathbf{0}$ is satisfied with non-zero scalars. Accordingly, $\mathbf{x}$ is a linear form of $\{\mathbf{x}_r\}$. The scalars in the representation of $\mathbf{x}$ can be calculated uniquely by projection as in $\mathbb{R}^3$.

Suppose now that the matrix $\mathbf{a}$ is not symmetric and denote by $\{\mathbf{u}_1, \ldots, \mathbf{u}_n\}$ and $\{\mathbf{v}_1, \ldots, \mathbf{v}_n\}$ its right and left eigenvectors. The right vectors are linearly independent since $\sum_{k=1}^{n} \alpha_k \, \mathbf{u}_k = \mathbf{0}$ implies $\alpha_k = 0$, $k = 1, \ldots, n$, by the orthogonality property. Multiply left the above expression by $\mathbf{v}_l^T$, $l = 1, \ldots, n$, to obtain $\sum_{k=1}^{n} \alpha_k \, \mathbf{v}_l^T \, \mathbf{u}_k = \alpha_l = 0$, $l = 1, \ldots, n$. Similar arguments hold for the set of left eigenvectors.

The sets of either left or right eigenvectors augmented with an arbitrary vector $\mathbf{x}$ are linearly dependent so that the relationships $\sum_{k=1}^{n} \alpha_k \, \mathbf{u}_k + \alpha_{n+1} \, \mathbf{x} = \mathbf{0}$ and $\sum_{k=1}^{n} \beta_k \, \mathbf{v}_k + \beta_{n+1} \, \mathbf{x} = \mathbf{0}$ hold with non-zero scalars. Accordingly, $\mathbf{x}$ can be represented by linear forms of the right or left eigenvectors whose components can be obtained uniquely by projection on the right or left eigenvectors.

Example C.10 The functions $\cos(\nu_k \, t)$ and $\sin(\nu_k \, t)$, $t \in [0, \tau]$, of Example C.3 define a basis $\mathcal{B}$ for the space of continuous periodic function on $[0, \tau]$ since they satisfy the conditions of Definition C.5. First, the elements of every finite subset of $\mathcal{B}$ are linearly independent since linear forms of these functions vanish only if their coefficients are zero. Second, the representation of f given by Eq. C.4 is unique by the orthogonality

$$
\begin{aligned}
&\int_0^\tau \sin(\nu_k \, t) \, \sin(\nu_l \, t) \, dt = \frac{\tau}{2} \, \delta_{kl}, \\
&\int_0^\tau \cos(\nu_k \, t) \, \cos(\nu_l \, t) \, dt = \frac{\tau}{2} \, \delta_{kl}, \\
&\int_0^\tau \cos(\nu_k \, t) \, \sin(\nu_l \, t) \, dt = 0
\end{aligned}
\tag{C.10}
$$

of the basis functions. For example, the inner product of $f(t)$ and $\cos(\nu_r \, t)$ is

$$
\begin{aligned}
\langle f(\cdot), \cos(\nu_r \, \cdot) \rangle = \frac{a_0}{2} \int_0^\tau \cos(\nu_r \, t) \, dt + \sum_{k=1}^{\infty} \left(a_k \int_0^\tau \cos(\nu_k \, t) \, \cos(\nu_r \, t) \, dt \right. \\
\left. + \, b_k \int_0^\tau \sin(\nu_k \, t) \, \cos(\nu_r \, t) \, dt \right) = \frac{\tau}{2} \, a_r,
\end{aligned}
\tag{C.11}
$$

where the integration was performed term-by-term. Technicalities on this operation can be found in [5] (Chap. 5).

Appendix D
Generalized Eigenvectors

The representation of the solution of MDOF systems as elements of the linear spaces spanned by modal shapes (proportional damping) or right/left eigenvectors (non-proportional damping) is possible if the eigenvalues are distinct so that the corresponding eigenvectors span the solution spaces. If one or more eigenvalues are multiple, the corresponding set of eigenvectors is insufficient to span the solution space. It has to be augmented by additional vectors, referred to as ***generalized eigenvectors***.

Definition 4 Consider an (n, n)-matrix $\mathbf{a}$ with an eigenvalue λ_1 of multiplicity k. The n-dimensional vector $\mathbf{x}_k$ defined by

$$\left(\mathbf{a} - \lambda_1\,\mathbf{I}\right)^{(k-1)}\mathbf{x}_j \neq \mathbf{0} \quad \text{and} \quad \left(\mathbf{a} - \lambda_1\,\mathbf{I}\right)^{(k)}\mathbf{x}_k = \mathbf{0} \tag{D.1}$$

is called ***generalized eigenvector*** or ***eigenvector of rank*** k of the (n, n)-matrix $\mathbf{a}$ associated with the eigenvalue λ_1.

Definition 5 Consider, as above, an (n, n)-matrix $\mathbf{a}$ with an eigenvalue λ_1 of multiplicity k. Define the vectors $\mathbf{x}_j$, $j = 1, \ldots, k-1$, by the following conditions:

$$\left(\mathbf{a} - \lambda_1\,\mathbf{I}\right)\mathbf{x}_k = \mathbf{x}_{k-1}, \left(\mathbf{a} - \lambda_1\,\mathbf{I}\right)^{(2)}\mathbf{x}_k = \mathbf{x}_{k-2}, \ldots, \left(\mathbf{a} - \lambda_1\,\mathbf{I}\right)^{(k-1)}\mathbf{x}_k = \mathbf{x}_1. \tag{D.2}$$

These vectors have two notable properties.

Property D.1 The vectors $\mathbf{x}_j$, $j = 1, \ldots, k-1$, are generalized eigenvectors of rank j, which means that $\left(\mathbf{a} - \lambda_1\,\mathbf{I}\right)^{(j-1)}\mathbf{x}_j \neq \mathbf{0}$ and $\left(\mathbf{a} - \lambda_1\,\mathbf{i}\right)^{(j)}\mathbf{x}_j = \mathbf{0}$.

Proof We have $\mathbf{x}_j = \left(\mathbf{a} - \lambda_1\,\mathbf{i}\right)^{(k-j)}\mathbf{x}_k$ from Eq. D.2 so that

M. D. Grigoriu, *Linear Dynamical Systems*,
https://doi.org/10.1007/978-3-030-64552-6

$$(\mathbf{a} - \lambda_1 \mathbf{I})^{(j)} \mathbf{x}_j = (\mathbf{a} - \lambda_1 \mathbf{I})^{(j)} (\mathbf{a} - \lambda_1 \mathbf{I})^{(k-j)} \mathbf{x}_k = (\mathbf{a} - \lambda_1 \mathbf{I})^{(k)} \mathbf{x}_k = \mathbf{0}$$

and

$$(\mathbf{a} - \lambda_1 \mathbf{I})^{(j-1)} \mathbf{x}_j = (\mathbf{a} - \lambda_1 \mathbf{I})^{(j-1)} (\mathbf{a} - \lambda_1 \mathbf{I})^{(k-j)} \mathbf{x}_k = (\mathbf{a} - \lambda_1 \mathbf{I})^{(k-1)} \mathbf{x}_k \neq \mathbf{0}$$

since $\mathbf{x}_k$ is a generalized vector of rank k.

Property D.2 The eigenvectors $\mathbf{x}_j$, $j = 1, \ldots, k$, are linearly independent.

Proof We need to show that $\alpha_1 \mathbf{x}_1 + \cdots + \alpha_j \mathbf{x}_j = \mathbf{0}$ implies $\alpha_1 = \cdots = \alpha_j = 0$. Since

$$\begin{aligned}(\mathbf{a} - \lambda_1 \mathbf{i})^{(j-1)} (\alpha_1 \mathbf{x}_1 + \cdots + \alpha_j \mathbf{x}_j) &= \alpha_1 (\mathbf{a} - \lambda_1 \mathbf{i})^{(j-2)} (\mathbf{a} - \lambda_1 \mathbf{I}) \mathbf{x}_1 \\ &+ \alpha_2 (\mathbf{a} - \lambda_1 \mathbf{i})^{(j-3)} (\mathbf{a} - \lambda_1 \mathbf{i})^{(2)} \mathbf{x}_2 \cdots + \alpha_j (\mathbf{a} - \lambda_1 \mathbf{I})^{(j-1)} \mathbf{x}_j = \mathbf{0},\end{aligned}$$

$(\mathbf{a} - \lambda_1 \mathbf{I}) \mathbf{x}_1 = \mathbf{0}$, $(\mathbf{a} - \lambda_1 \mathbf{I})^{(2)} \mathbf{x}_2 = \mathbf{0}$, and so on, while $(\mathbf{a} - \lambda_1 \mathbf{i})^{(j-1)} \mathbf{x}_j \neq \mathbf{0}$, we have $\alpha_j = 0$. Similar arguments hold for all other values of $j = 1, \ldots, k-1$.

These properties show that we can construct a set of k generalized eigenvectors for each eigenvalue of multiplicity k and that these vectors span a k-dimensional subset of $\mathbb{R}^k$. They also show that the displacement vectors for MDOF systems can be represented by projections on the generalized eigenvectors, rather than eigenvectors, for systems with multiple eigenvalues.

Example D.1 Consider the eigenvalue problem for the (3,3)-matrix

$$\mathbf{a} = \begin{bmatrix} 1 & 1 & 2 \\ 0 & 1 & 3 \\ 0 & 0 & 2 \end{bmatrix}.$$

The eigenvalues are the roots of the polynomial $\det(\mathbf{a} - \lambda \mathbf{I})$ in λ, i.e., the solutions of $(\lambda - 1)^2 (\lambda - 2) = 0$, so that $\lambda_1 = 1$ and $\lambda_2 = 2$ are double and simple eigenvalues. Denote by $\mathbf{x}_1$ and $\mathbf{x}_3$ the eigenvectors of λ_1 and λ_2. Their components are $(1, 0, 0)$ and $(5, 3, 1)$. Clearly, $\mathbf{x}_1$ and $\mathbf{x}_3$ do not span $\mathbb{R}^3$. We need an additional vector to span this space. This vector, denoted by $\mathbf{x}_2$ and called generalized eigenvector, is the non-trivial solution of

$$(\mathbf{a} - \lambda_1 \mathbf{I})^{(2)} \mathbf{x}_2 = (\mathbf{a} - \lambda_1 \mathbf{I})\big[(\mathbf{a} - \lambda_1 \mathbf{I}) \mathbf{x}_2\big] = \mathbf{0},$$

such that $(\mathbf{a} - \lambda_1 \mathbf{I})\mathbf{x}_2 = \mathbf{x}_1$ (see Eq. D.2) or, equivalently, $\mathbf{a}\,\mathbf{x}_2 = \lambda_1 \mathbf{x}_2 + \mathbf{x}_1$. The latter equation gives the conditions

$$x_{2,1} + x_{2,2} + 2\,x_{2,3} = x_{2,1} + 1$$

$$x_{2,2} + 3\,x_{2,3} = x_{2,2}$$
$$2\,x_{2,3} = x_{2,3}$$

for the components $\{x_{i,k}\}$ of $\mathbf{x}_i$ since $x_{1,1} = 1$ and $x_{1,2} = x_{1,3} = 0$. The last equation implies $x_{2,3} = 0$, a result that is consistent with the second equation. The first equation with $x_{2,3} = 0$ gives $x_{2,2} = 1$ so that the components of $\mathbf{x}_2$ are $(0, 1, 0)$. The vectors $(\mathbf{x}_1, \mathbf{x}_2, \mathbf{x}_3)$ span $\mathbb{R}^3$. The eigenvectors $\mathbf{x}_1$ and $\mathbf{x}_2$ are aligned with the first coordinates of $\mathbb{R}^3$, while $\mathbf{x}_3$ has non-zero projections on coordinates of this space.

References

1. C.M. Bender, S.A. Orszag, *Advanced Mathematical Methods for Scientists and Engineers* (McGraw-Hill, New York, 1978)
2. R.L. Brabenec, *Introduction to Real Analysis* (PWS-KENT Publishing Company, Boston, 1990)
3. D.E. Goldberg, *The Design of Innovation: Lessons from and for Competent Algorithms* (Kluwer Academic Publishers, Norwell, 2002)
4. D.A. Sánchez, *Ordinary Differential Equations and Stability Theory: An Introduction* (Dover Publications Inc., Mineola, 1968)
5. G.P. Tolstov, *Fourier Series* (Dover Publications Inc., New York, 1962)

Index

M. D. Grigoriu, *Linear Dynamical Systems*,
https://doi.org/10.1007/978-3-030-64552-6

Printed in the United States
by Baker & Taylor Publisher Services